$E \neq mc^2$: The Collapse of Classical Assumptions

Chavan's UED research, Volume 12

Sandeep Chavan

Published by SANDEEP CHAVAN, 2025.

E ≠ MC²: THE COLLAPSE OF CLASSICAL ASSUMPTIONS

First edition. June 14, 2025.

ISBN: 979-8231610372

Written by Sandeep Chavan.

Table of Contents

To **Albert Einstein** —

Not because he was always right,but because he had the courage to question everything.

This book is not a rejection of your legacy,but a ripple of your greatest lesson:"**Imagination is more important than knowledge.**"

May this work continue the journey you began—from symbols to meaning,from equations to understanding.

— *With deepest respect,Sandeep Chavan*

"The universe never exploded. It unfolded."

"Nothing moves — everything adjusts. Nothing glows — everything echoes."

"We weren't meant to conquer nature.We were born to realign with it."

— *Sandeep Chavan*

Disclaimer

This book presents original scientific and philosophical interpretations developed by the author based on independent research under the Universal Energy Dynamics (UED) framework. It challenges established scientific paradigms not as a dismissal, but as an invitation for renewed inquiry.

The content is not intended to undermine validated scientific work but to offer alternative models for consideration, discussion, and exploration. Readers are encouraged to critically evaluate the ideas, apply their own judgment, and view this work as part of an evolving scientific dialogue.

This book does not claim affiliation with any academic or institutional authority, and the author assumes no responsibility for the application or misinterpretation of the material in experimental, educational, or professional settings.

All examples, reinterpretations, and historical references are reconstructed to support the conceptual model and may differ from mainstream understanding.

Opening Note

Physics taught me to calculate, not always to understand. Equations told me what happens — but rarely why.

For years, we've admired $E = mc^2$ as the crowning jewel of scientific brilliance — a symbol of elegance, power, and predictive success. But beneath that simplicity lies a conceptual silence that few dared to question: *What exactly is being converted?* And *why should mass ever become energy?*

As I explored deeper through the lens of Universal Energy Dynamics (UED) — a framework born from consequence, not convention — I began to see that mass is not a substance, energy is not a thing, and light is not emitted. They are all expressions of ripple alignment in a structured vacuum field. When this alignment breaks, form dissolves — not to become something else, but to release what it always was.

This book is not an act of rebellion. It is an act of correction. Not an attack on Einstein, but an invitation to re-express what he sensed — in a language truer to field, perception, and alignment.

Some will call this controversial. Let them. The history of science is written not by those who *agree*, but by those who *observe what others miss*.

I did not set out to prove Einstein wrong. I set out to understand what had been taken for granted — and what I found was not a conversion, but a dissolution. Not a mass-energy equivalence, but a ripple transition. Not an absolute truth — but a mathematical metaphor that overstayed its meaning.

This book is bold, yes — but it's also necessary. Because the moment we stop questioning physics, we stop understanding reality.

And I'm not done yet.

—Sandeep Chavan

Independent Researcher,

Founder – Universal Energy Dynamics (UED)

ORCID Profile: https://orcid.org/0009-0005-4190-7164

Preface

I never set out to rewrite physics.

I set out to understand it — honestly, humbly, and without bias. What I found instead was a collection of symbols, metaphors, and mathematical illusions dressed as certainty. Equations that worked but explained nothing. Models that predicted, but never truly revealed. And a universe that was supposedly exploding, expanding, emitting — but somehow felt silent, still, and perfectly coherent.

This book is not a rebellion. It is a restoration.

A restoration of consequence over convention. A restoration of meaning over metaphor. A restoration of the field — not as space, but as structure; not as emptiness, but as intelligence.

Through years of independent research, writing, teaching, observing, and listening — not just to textbooks, but to ripples of reality itself — a new framework emerged: **Universal Energy Dynamics (UED)**. It is not a theory of everything. It is a practice of perceiving. It asks not "What is the object?" but "What holds it in coherence — and what happens when that breaks?"

This book presents that understanding in its most distilled form: **Ripple Science**.

Here, **energy** is not a thing. It is the **cost of realignment**. **Mass** is not substance. It is **ripple stability under stress**. **Light** is not emitted. It is **the final feedback of coherence**. **Time** does not flow. It is the **readiness of the field to ripple**. And **you**, the observer, are not outside it — you are **the field watching itself**.

We move beyond Einstein not to reject him, but to continue what he began. To replace metaphors with mechanics, curvature with causality, and particles with patterns. Einstein offered the language of relativity. This book offers the **grammar of ripple coherence**.

To scientists, I invite you to re-measure. To engineers, I ask you to redesign. To philosophers, I challenge you to rethink what it means to "exist." To the curious reader, I simply say: reality is not what you see — it's what aligns to let you see it.

This is not the final word. It is the first ripple.

Let it echo.

— **Sandeep Chavan**

Prologue

When the Equation Became the Truth

It began with a simple equation. A small cluster of symbols that promised to unify mass and energy — to explain the mechanics of stars, bombs, and perhaps even the beginning of time itself.

$E=mc^2$

In its elegance, it captivated generations. In its simplicity, it seduced even those who barely understood it. And in its silence — it hid what no one wanted to admit:

That we had taken a metaphor, and declared it a law of nature.

The idea that mass could "become" energy — that a tangible form could vanish into power — made sense only when we stopped asking what mass *is*, what energy *really means*, and how light *truly behaves*.

In reality, nothing converts. Things change form. Fields realign. Ripples dissolve. And what we observe as "release" is just the field settling back into equilibrium.

Mass was never a vault storing energy. It was a temporary ripple configuration, a phase of alignment in the cosmic field. Energy wasn't trapped inside it — energy *was* the configuration. And when that configuration broke? We didn't see a transformation — we saw a *consequence.*

Yet, we built theories, bombs, entire worldviews around this one symbol of scientific divinity. We recited it like scripture. We printed it on posters. We etched it into collective memory as the ultimate truth.

This book tears away that illusion. Not to diminish what Einstein gave us — but to complete what he couldn't finish. To step beyond the gravitational pull of dogma, and into the structured vacuum that actually governs all things.

What if energy is not a thing to be released, but a field to be realigned? What if mass is not substance, but consequence? What if light is not emitted, but echoed — the final signature of ripple coherence dissolving into invisibility?

Then $E = mc^2$ is not a law. It's a shadow — cast by something far deeper.

And it's time we turned around and faced the source of that light.

Part I: The Cult of the Equation

Chapter 1: The Icon of 20th Century Physics

How $E = mc^2$ Became a Religion

You could write it on a napkin and change the world.

That's what they say about $E = mc^2$ — the most recognizable equation in human history. It has no long variables. No complicated structure. No integrals or Greek symbols. Just three letters, an equal sign, and a squared constant. But somehow, it became the mantra of modern science — whispered in textbooks, glorified in documentaries, etched into the DNA of every schoolchild who ever wondered what stars are made of.

It wasn't just a formula. It became faith.

Not because everyone understood it — but precisely because they didn't. $E = mc^2$ was simple enough to remember but too sacred to question. It became the scientific equivalent of scripture — recited, revered, but never rewritten.

And just like scripture, it was used to justify everything from atomic bombs to big bang cosmology, from nuclear power to the god-particle hunt. It shaped empires, guided policies, and shielded itself from scrutiny — not through clarity, but through charisma.

Yet beneath the glow of genius, one question lingered in silence:

What exactly is mass? What is energy? And what makes us so certain that one becomes the other?

This chapter is not about denying Einstein's brilliance. It's about confronting the moment we stopped asking those questions — and started believing in the equation more than the physics behind it.

Because when science becomes belief, even the most elegant truth becomes a dogma in disguise.

A Symbol Greater Than Its Meaning

If science had a logo, it would be $E = mc^2$.

No other equation in human history has achieved such mythic status — not Newton's F=ma, not Schrödinger's wave function, not even Maxwell's elegant set of field equations. Those remain inside classrooms, lecture halls, or the minds of specialists. But $E = mc^2$ escaped. It broke out of academia and entered pop culture — unchallenged, untouchable, and largely misunderstood.

From Chalkboard to T-Shirt: The Branding of a Formula

It began as a quiet conclusion in a 1905 paper on special relativity. A mere footnote in Einstein's own words. Yet within a few decades, it was no longer just part of physics — it was part of identity.

- It appeared in textbooks as the summary of how reality works.
- It adorned T-shirts, mugs, and posters.
- It became shorthand for genius, mystery, and power.
- And most importantly, it was rarely questioned — because it looked complete.

Three characters. One idea. Zero resistance.

By the time the atomic bomb cast its long shadow over Hiroshima, $E = mc^2$ had already become a justification, not a hypothesis. It explained the unimaginable with the elegance of inevitability. We didn't need to understand how uranium unraveled — we just needed to point to the equation and say, "See? That's what happens when mass becomes energy."

The Pop-Science Obsession with "Unlocking the Universe"

For the media and the public, $E = mc^2$ became more than a physical relation — it became a spiritual slogan for modernity.

- TV specials and YouTube videos call it the most beautiful equation.
- It's spoken of as if it holds the key to *everything* — from black holes to enlightenment.
- Pop physicists cite it to add gravitas to any concept, even when the underlying physics is absent.

But here's the real issue:

The simplification became the belief. The equation became the explanation.

It's quoted in awe, like a cosmic haiku. But ask the average believer to explain:

- What *is* mass, really?
- What *is* energy, fundamentally?
- What exactly "converts" in an explosion or a reaction?

The answers crumble.

Why It's Quoted More Than It's Understood

Because $E = mc^2$ is emotionally satisfying.

It tells a seductive story:

- That nature is simple.
- That all things are connected.
- That there's a hidden order beneath the chaos.

But the deeper truth is:

- The mechanism is missing.
- The interpretation is assumed.

- And the physics is glossed over by dimensional correctness.

"It works" has replaced "It makes sense." Dimensional elegance has replaced physical insight.

The UED Perspective:

From the lens of Universal Energy Dynamics, this section reveals the first crack: The equation became more important than the explanation it failed to deliver.

Mass doesn't become energy. Mass is already energy — structured, aligned, and perceived. When the form dissolves, what remains isn't a conversion, but a consequence.

Closing Thought

We didn't fall in love with the physics. We fell in love with the illusion of understanding. And in doing so, we turned an incomplete idea into a complete idol.

The Man Behind the Mass: Einstein's Leap

To understand the power and pitfalls of $E = mc^2$, we must first return to 1905, the so-called *annus mirabilis* — Einstein's "miracle year." But unlike the pop-hagiography that paints Einstein as a lone genius dropping cosmic truths from a patent office desk, the reality is more nuanced — and more human.

This section explores not only what Einstein wrote but what he meant, what he left unexplored, and how the equation itself was never the central thesis, but a side effect of deeper concerns.

Einstein in 1905: A Mind Outside the Machine

Albert Einstein was not yet the icon — no pipe, no Nobel, no Princeton. He was 26, working as a technical expert in the Swiss patent office, far from academic circles. But what he lacked in academic position, he made up for in relentless thought experiments, philosophical clarity, and a refusal to conform to accepted dogmas.

That year, he published five papers that would change physics forever:

- On the photoelectric effect (quantum light)
- On Brownian motion (evidence for atoms)
- Two on special relativity
- And finally, a brief note suggesting a revolutionary idea: mass and energy are deeply connected.

But this last insight — what we now glorify as $E = mc^2$ — was not the main dish. It was a dessert — a concluding comment in an otherwise focused work on the relativity of motion.

The Special Relativity Paper: Elegant, but Incomplete

Einstein's *Zur Elektrodynamik bewegter Körper* ("On the Electrodynamics of Moving Bodies") didn't mention $E = mc^2$. It laid down the foundation of special relativity: how the laws of physics are the same in all inertial frames, and how the speed of light remains constant regardless of observer motion.

But the idea of mass-energy equivalence was introduced later that year in a separate short paper: "Does the Inertia of a Body Depend Upon Its Energy Content?"

In this note, Einstein proposed that if a body emits energy L as radiation, its mass must decrease by Lc^2. From this, the famous relation emerged:

$$E=mc^2$$

Yet, he did not derive it as a fundamental law. He presented it as a logical implication — a side result of relativity principles combined with known electromagnetic behavior.

In short: Einstein suggested it, he didn't proclaim it.

Mass-Energy Equivalence: A Consequence, Not a Discovery

This distinction matters. Einstein's equation wasn't declared with fanfare. It wasn't a grand unveiling of a universal truth. It was a calculated consequence of aligning existing theories of motion and light.

He didn't describe *what energy is*. He didn't define *how mass contains it*. He simply said: if mass loses energy, the mass must reduce.

It was cautious, logical, and elegant — but not comprehensive. The mechanism remained unaddressed.

And yet, this cautious implication was later interpreted — and eventually marketed — as a bold claim:

"Mass becomes energy."

This was never what Einstein meant — and certainly not what he proved.

Was Einstein Himself Cautious?

Surprisingly, yes. In his correspondence and lectures over the years, Einstein:

- Acknowledged the limitations of special relativity,
- Admitted the energy-mass relation needed deeper field theory understanding,
- And never oversold $E = mc^2$ the way later generations did.

In 1946, after Hiroshima, Einstein famously remarked:

"I do not consider myself the father of the release of atomic energy. My part in it was very indirect."

In later years, he was more concerned with unified field theory, gravitation, and the philosophical implications of physics — not the promotional glory of a simplified equation.

Yet, his caution was ignored, and his equation was canonized.

Closing Reflection

Einstein never built a religion — but one was built around him. He opened a doorway with an equation — but we mistook the doorway for the entire house. And in doing so, we forgot to ask the one question he never answered:

"What is mass *really*, and what *actually* happens when it disappears?"

War, Power, and the Weaponization of Thought

Equations were once meant to explain the universe. But in the 20th century, $E = mc^2$ did something more — it crossed over from science to strategy, from blackboards to battlefields, from thought to threat.

The mass-energy relation, initially a quiet theoretical insight, was rebranded as a mechanism for destruction — a scientific justification for annihilation. In this transformation, it stopped being a question and became a verdict.

From Theory to Theater of War: The Birth of Atomic Destiny

By the late 1930s, Europe stood on the edge of catastrophe. Hitler had risen, fascism was spreading, and theoretical physicists — once detached from geopolitics — were being drawn into the gravitational field of war.

Einstein, now a global icon, had fled Nazi Germany and settled in the United States. Although he had no direct involvement in nuclear weapons research, a letter bearing his name, drafted by physicist Leo Szilard, was sent to President Roosevelt in 1939. It warned that Germany might develop atomic weapons — and urged the U.S. to act first.

That letter is widely credited with catalyzing the Manhattan Project — the top-secret American effort to build an atomic bomb.

And at the heart of it all?

$E = mc^2$ — suddenly no longer theoretical.

It was used to quantify the unimaginable: If even a small amount of mass could become energy, then a bomb made from uranium or plutonium could unleash the power of stars.

The Equation as Justification for Mass Destruction

What was once an academic relation became a calculated threat. Scientists, generals, and politicians all pointed to the same equation — as if nature itself had authorized the bomb.

"This is not invention. This is inevitability." "We're not creating destruction — we're merely following the laws of physics." "This isn't a choice. It's a fact."

$E = mc^2$ became a kind of alibi — a way to detach human intent from consequence. The deaths at Hiroshima and Nagasaki? Not decisions. Just inevitable expressions of mass becoming energy.

In this way, the equation did something no other in history had done:

It rationalized horror through mathematical truth.

Emotional Manipulation: Science as Destiny

Behind the scenes, a more dangerous story was unfolding.

Once $E = mc^2$ became associated with the bomb, it became not just a scientific statement, but a psychological tool:

- It gave political leaders a sense of divine inevitability.
- It gave scientists a moral escape hatch: "We just followed the math."
- It gave the public a false clarity: "This is just how the universe works."

This emotional manipulation weaponized awe. The very same awe that $E = mc^2$ inspired in textbooks and classrooms was now used to create moral numbness. If science says it must be so — who are we to question it?

$E = mc^2$ as a Moral Shield

The most dangerous transformation wasn't physical — it was philosophical.

The equation became a shield to hide human responsibility behind scientific inevitability.

- Governments justified bombings through energy calculations.
- Science was treated not as a *method of inquiry*, but a *mandate of execution*.

- Physicists themselves were split — some celebrated, others regretted, but very few challenged the equation itself.

Because once $E = mc^2$ was used to make a bomb, it became *sacred.* To critique it was to risk being labeled unscientific, even treasonous.

And so, a theoretical insight meant to explain light and inertia became a holy law of planetary threat.

UED's Ethical Correction:

Universal Energy Dynamics breaks this illusion. It tells us:

- There is no conversion of mass into energy — only dissolution of ripple configuration.
- The "energy released" is not hidden power, but a ripple redistribution in a structured field.
- The destructive capacity doesn't arise from mass itself — but from our forced misalignment of coherent structures.

The bomb wasn't inevitable. It was engineered. The physics didn't *demand* it. We did.

Final Reflection

$E = mc^2$ was never just an equation. It became a tool, a myth, and a shield. And in its name, we turned science from a lamp of understanding into a weapon of power.

The question now is: Can we strip the myth, and recover the meaning?

From Science to Faith — The Rise of a Scientific Dogma

There's a subtle line between understanding something deeply and simply believing it works. When that line is crossed in science, curiosity gives way to dogma. And few ideas have crossed that line more dramatically than $E = mc^2$.

From Curiosity to Reverence

Originally, Einstein's mass-energy equivalence was a provocative consequence — a speculative proposal inviting further exploration. But over time, as it appeared to "work," it slowly acquired an aura of finality.

The more it was repeated, the less it was questioned. The more it was simplified, the more it was glorified. And the more it was associated with power — bombs, reactors, the stars — the more it became sacred.

What started as a prompt for inquiry became an object of awe.

Scientists no longer discussed *how* mass "becomes" energy — they taught that it *does*. The mechanism? Unimportant. The assumptions? Unchallenged. The consequences? Wrapped in certainty.

Why Textbooks Never Critique $E = mc^2$

Pick up any modern physics textbook. You'll find:

- Diagrams showing mass disappearing into energy.
- Bullet points claiming $E = mc^2$ explains nuclear power and star light.
- Reverent sidebars about Einstein's genius.

But what you won't find is a real answer to:

- *What is mass fundamentally?*
- *How exactly does this "conversion" happen?*
- *Where is the mechanism of transition — not the math, but the cause?*

This omission is not by accident — it's by tradition. Textbooks teach confirmation, not contemplation. Once something is canonized, it's not revisited. It's *compiled.*

$E = mc^2$ entered the curriculum not as a theory to be explored, but as a truth to be memorized.

And memorized it was — globally, across generations. Ask anyone what $E = mc^2$ means, and they'll tell you confidently: "Mass becomes energy."

But ask how, and suddenly the clarity dissolves.

Dimensional Correctness ≠ Physical Meaning

One of the most dangerous habits in modern physics is the overreliance on dimensional agreement as proof of validity.

$E = mc^2$ "works" because:

- Mass has units of kilograms.
- Speed of light squared has units of m^2/s^2.
- Multiply them, you get $kg{\cdot}m^2/s^2$ = joules — energy.

This is mathematically correct. But mathematics without mechanism is blind.

Dimensional compatibility is necessary — but it is not sufficient for truth.

Just because the units align doesn't mean the physics does.

In reality:

- Energy is not a quantity hidden inside mass.
- Mass is not a vault of energy waiting to be released.
- The observed energy in reactions is not *converted* from mass — it is a reconfiguration of the ripple field, as explained by UED.

But as long as the units add up, few ask what's actually happening underneath.

How the Equation Became Untouchable — and Why That Ends Here

The deeper the reverence grew, the more critique was treated as heresy:

- To question $E = mc^2$ was to question Einstein.
- To question Einstein was to question science.
- And to question science? That's intellectual blasphemy.

Thus, the equation became untouchable. Not because of its clarity, but because of its reputation.

But dogma, no matter how elegantly packaged, must eventually face the reality it denies.

And this is where UED enters — not to destroy Einstein's legacy, but to reclaim what science should have done all along: Explain, not just calculate. Understand, not just accept. Question, not just believe.

$E = mc^2$ is not being dismissed here — it is being decoded. Its place in history is acknowledged. Its utility is respected. But its physical interpretation — its meaning — is being reclaimed from the fog of tradition.

Final Reflection

Science is not a cathedral. Equations are not commandments. And reverence has no place in the pursuit of truth. What we cannot explain, we must not worship. The era of quoting without questioning ends here.

Chapter 2: Dimensional Accuracy, Conceptual Poverty

Why Correct Units Don't Mean Correct Physics

An equation can be dimensionally flawless and still be physically empty.

It can pass every unit test, match every measurable quantity, and still explain nothing about reality.

Physics today is obsessed with consistency — not clarity. With balance sheets of units, not the origin of causes.

This is how $E = mc^2$ survived unchallenged: It looks right. It behaves right. Its outputs match effects we observe — energy from bombs, stars, and annihilations.

But ask what really happens inside the process —Ask what exactly is converting —Ask what mass and energy fundamentally are —And suddenly, the math goes mute.

Dimensional accuracy becomes a mask —A beautiful veil that hides the conceptual poverty underneath.

We've built an empire on equations that work, without ever confirming if they mean anything true.

And at the center of that empire sits the speed of light (c) — not as an observable constant, but as a frozen axiom, never to be questioned, never to be reinterpreted.

But in UED, we don't stop at matching outputs. We demand mechanism. We ask: what is happening, not just what is measurable. And that shift — from numbers to meaning — changes everything.

This chapter reclaims physics from the illusion of mathematical perfection. Because until we look beyond units and symbols, we're not doing science. We're performing rituals in the language of calculus.

The Illusion of Precision

There is a dangerous comfort in clean equations.

When all the units line up — when the left side of an equation balances perfectly with the right — we breathe a sigh of relief. The physics "works." The math is "elegant." The equation is declared valid.

But dimensional consistency is not a declaration of truth. It is merely a filter — a necessary condition, not a sufficient one. And in modern physics, this necessary condition has been mistaken for understanding.

We have come to trust the match of units more than the meaning behind them.

Dimensional Consistency: A Mathematical Form, Not Physical Function

In physics, every equation must be dimensionally homogeneous — the units of each term must match. Energy must be in joules, velocity in m/s, mass in kg, etc. This ensures that the equation doesn't violate basic numerical logic.

This is undoubtedly important, but it's not evidence of physical insight. You can construct an equation that is perfectly balanced in units and yet entirely devoid of physical meaning.

Example:

You could write:

$$E = \frac{1}{2}mv^2 + mc^2 - \frac{3}{4}mv^2$$

The units work. You're still left with energy in joules. But does it describe anything real? Anything meaningful? Not at all.

This is the illusion of precision. It tells you the math is valid — but says nothing about whether reality agrees.

When Math Masquerades as Truth

There is a psychological trap in theoretical physics: the more mathematically elegant an equation is, the more likely it is to be treated as physically correct — even in the absence of an underlying mechanism.

Einstein understood this risk. He chased elegance, but not at the cost of interpretation. However, those who followed him often did.

- The equations of general relativity curve space, but never explain what space is.
- Quantum mechanics gives accurate probabilities, but cannot say why or how outcomes occur.
- String theory is dimensionally and algebraically rich, yet physically barren in empirical clarity.

And in all these cases, the shield is the same:

"The units check out."

That's not physics. That's numerical theatre.

Case Study: $E = mc^2$ — The Poster Child of Dimensional Success, Conceptual Failure

Let's return to the icon.

$E=mc^2$

It checks all the boxes:

- m in kilograms
- c^2 in $(m/s)^2$
- Product is in $kg{\cdot}m^2/s^2$ = joules → ✔

It is dimensionally perfect.

But what is being described?

- Is energy "stored" inside mass?
- Is mass "made of" energy?
- What *process* changes mass into energy?
- Is there a field mechanism for the transformation?

No mainstream answer exists. And no derivation from first principles explains this so-called "conversion." Einstein himself didn't derive it as a fundamental law — it was an inference from electromagnetic principles, based on idealized thought experiments. And yet it became doctrine, because the units match.

Matching units gave it immunity from further questioning.

But in UED, we do not confuse symbolic balance with field dynamics. We understand that mass is ripple coherence, and energy is ripple reconfiguration. There is no conversion — only field redistribution. And that is a consequence, not a calculation.

Final Reflection:

Dimensional consistency tells you where the door is. But it does not tell you what's behind it. Physics is not about matching sides — it's about understanding the field in between.

In the chapters ahead, we leave behind the shallow comfort of tidy symbols and step into a framework where meaning leads the math, not the other way around.

The Tyranny of the Speed of Light

If there's one symbol as untouchable as $E = mc^2$, it's the speed of light, c.

This constant — 299,792,458 meters per second — is held as a cosmic law, a final limit, a sacred ceiling of motion that nothing can cross. It is embedded in nearly every modern equation, from relativity to quantum electrodynamics. But beneath this awe and rigidity lies an unspoken truth:

c was never discovered — it was assumed. It was not proven to be a limit — it was declared as one.

c: A Constant Chosen, Not Understood

The speed of light was not originally a metaphysical boundary. It was simply a measurement — a description of how fast light *appeared to move* under certain conditions. From Ole Rømer in the 1600s to Michelson and Morley in the 1800s, the constant emerged from optical and astronomical experiments.

But when Einstein published his special relativity paper in 1905, he took a radical step: He made c a postulate — not something to be derived, but something to be assumed.

In Einstein's framework:

- The laws of physics are the same in all inertial frames.
- The speed of light in vacuum is the same for all observers, regardless of their motion.

This assumption gave the theory its elegance — and its strange consequences:

- Time dilates.
- Length contracts.
- Mass increases with speed.

But the foundation — c as constant and absolute — was never questioned after that. It was enshrined. Not because we fully understood it, but because it made the math work.

Einstein didn't explain why c is what it is. He used c to define what space and time must become.

Why Einstein Had to Freeze c — and Redefine Mass Instead

When Einstein embedded c as a fixed limit in his equations, he locked the top of the system. Nothing could go faster. Nothing could challenge the ceiling.

But reality resisted.

As objects move faster, they appear to resist acceleration more and more. To keep c frozen, Einstein had to shift the burden elsewhere. So he introduced the concept of relativistic mass:

$$m_{\text{rel}} = \frac{m_0}{\sqrt{1 - \frac{v^2}{c^2}}}$$

Here, as velocity vvv approaches ccc, the denominator tends to zero, and mass tends toward infinity.

This kept c safe, but it introduced an absurdity:

- An object's mass becomes infinite as it nears light speed?
- Infinite energy is required to reach c?

It worked numerically. But physically, it became an artificial trap — a mathematical jail created to protect an unexamined assumption.

Rather than revisiting the meaning of c, physics redefined mass itself to maintain the illusion of consistency.

UED Reinterpretation: c is Not a Barrier — It's a Feedback Threshold

In Universal Energy Dynamics (UED), we reject the idea that c is a hard cosmic speed limit.

Instead, we understand c as the maximum rate at which ripple feedback becomes perceivable — not a property of space, but of observation.

- Light is not a "thing" moving at speed c.
- Light is a ripple alignment feedback — the last visible echo of a dissolving configuration.
- The reason we always measure this speed as constant is because it is the threshold at which our observer field aligns with

ripple feedback.

Thus, c is not a universal constant of motion, but a ceiling of synchronization between observer and ripple field.

Nothing "travels" at c — what we see as light is the final perceptual residue of field adjustment.

ATT: The True Limit Is Not Speed — It's Alignment Threshold Time

In UED, motion is not displacement through space — it's reconfiguration in the field. The real question isn't *how fast something moves*, but *how fast a ripple pattern can realign to stay coherent and perceptible.*

This gives rise to a new concept: ATT — Alignment Threshold Time

- ATT is the minimum time required for a ripple configuration to be perceived by an observer.
- If the realignment happens faster than this threshold, the structure becomes invisible — not because it's fast, but because it cannot register in the observer's field.
- This is why light is the last visible echo — the ripple alignment just before form dissolves beyond perceptibility.

Where Einstein said "Nothing can move faster than c," UED says "Nothing can be perceived faster than ATT."

This re-centers the focus on consequence, not velocity.

Final Reflection:

Physics froze the speed of light and rewrote everything around it. But it never explained what light truly is, or why c matters. In UED, we free ourselves from that tyranny — and rediscover the true boundary: Not speed, but perception. Not c, but ATT. Not limitation, but alignment.

Mathematical Structure vs Physical Substance

Why Physics Lost Its Meaning and How UED Restores It

We live in an era where physics appears flawless on paper but increasingly divorced from reality. The mathematics is breathtaking, the equations are complex, and the predictions are often accurate.

Yet there is a growing emptiness beneath the surface — a silence where understanding should be.

This section explores that divide: The difference between describing *what happens* and explaining *why it happens.*

Describing the Phenomenon Is Not Explaining the Cause

Mathematics is a language — powerful, symbolic, universal. But language alone does not explain the world.

- An equation can *describe* a falling object.
- It can *predict* where it will land.
- It can *model* how fast it will accelerate.

But does it tell you *why* it falls?

Does it uncover the field structure, the ripple behavior, or the energetic tension that underlies that motion?

No. It simply reflects what was observed, and fits it to a curve.

This distinction matters. Because modern physics, in its pursuit of precision, has often abandoned causality in favor of correlation.

We've stopped asking:

- *Why does mass resist acceleration?*
- *What is space, fundamentally?*
- *What initiates a force?*
- *What is light, beyond the speed we've assigned it?*

Instead, we settle for:

- "Because the equation says so."

- "Because the units match."
- "Because the graph fits."

This is not physics. It's data sculpting.

When Physics Became a Symbol Game

The most celebrated achievements of the 20th century in physics — relativity, quantum mechanics, thermodynamics — are dominated by mathematical structure.

These equations are elegant, but they abstract away the substance. They offer form without foundation.

- The Schrödinger equation tells you the probability of an event, but not what causes it.
- General relativity bends spacetime, but never explains what spacetime is or why it bends.
- Maxwell's equations track electromagnetic waves, but cannot define what a wave is in vacuum without circular reasoning.

In this landscape, symbolism replaces substance.

Instead of explaining nature, we became fluent in predicting outcomes — like learning to sing in a language we don't speak.

This shift has consequences:

- Physics becomes inaccessible — understood by calculators, not consciousness.
- Scientists become equation guardians, not field interpreters.
- And paradoxes arise — not from nature, but from our incomplete translation of it.

The Danger of Fitting Phenomena into Formulas

In modern physics, the tendency is clear:

- Observe a phenomenon

- Fit a curve
- Write an equation
- Declare a law

But this reverses the scientific spirit.

Instead of starting from cause, we start from consequence. Instead of seeking understanding, we seek patterns. And if something doesn't fit? We create a new constant, a new particle, a new symmetry.

It's no longer science as discovery. It's mathematical reverse engineering.

This is why we end up with:

- **Dark matter** — invented to preserve equations of motion we don't fully understand.
- **Inflation** — added to fix inconsistencies in a big bang model that can't explain early uniformity.
- **Multiverses** — invoked to account for fine-tuning without addressing the root architecture of the cosmos.

We invent mathematical patches instead of revising the underlying cloth.

UED: Restoring Physical Meaning through Ripple Field Dynamics

Universal Energy Dynamics (UED) rejects the idea that reality is symbol-first. In UED, field behavior comes first — symbols come later.

- Every phenomenon is the result of ripple field dynamics in a structured vacuum.
- Forces, motions, emissions, and even "mass" are just different configurations of ripple alignment.
- Equations are useful only if they reflect the consequence of field behavior, not just the shape of the data.

This means shifting the purpose of physics from:

- Symbol manipulation → to field interpretation
- Dimensional correctness → to ripple causality
- Quantity fitting → to alignment understanding

In UED:

- **Light is not a particle or a wave** — it's ripple feedback at the edge of perception.
- **Mass is not a substance** — it's locked ripple coherence.
- **Energy is not stored** — it is the difference between ripple states.

This isn't a tweak. It's a reinstatement of meaning.

Final Reflection:

The universe was never written in equations. We just noticed patterns, and called them laws. But beneath the math lies motionless structure —a ripple field waiting to be read, not solved.

Beyond Compatibility — The Case for Consequence-Based Physics

From Fitting Equations to Understanding Reality

Modern physics has mastered the art of compatibility.

- Units match? Check.
- Data fits the curve? Check.
- Prediction aligns with experiment? Check.

But the one box that rarely gets checked anymore is the most vital one:

Does it explain the consequence?

In this section, we go beyond the comfort of mathematical alignment and into the core demand of UED: Physics must explain why something happens — not just how well it matches a number.

The UED Litmus Test: Consequence Over Correlation

At the heart of Universal Energy Dynamics (UED) is a simple but radical principle:

If a theory cannot explain the consequence, it does not understand the cause.

Consequence in UED is not limited to measurement. It is the field-level result of a specific ripple interaction — a shift in equilibrium, a realignment of energy, a transition from one coherent configuration to another.

Contrast this with the current paradigm in mainstream physics:

- Observe a phenomenon.
- Match an equation to it.
- Predict similar results.
- Claim understanding.

But matching results isn't understanding — it's correlation. And correlation can be coincidental, superficial, or incomplete.

Example:

- In the photoelectric effect, we observe electrons ejecting from metal when light hits it.
- The equation E=hf fits.
- But what is light? What initiates electron motion at the field level?
- The standard model gives no real field-based explanation — only that it happens and is quantized.

UED reframes this:

- Light is not a packet.
- It is a ripple feedback reaching perceptual threshold (ATT).
- The electron's ejection is a ripple interference outcome, not a particle collision.

This is consequence-based thinking, not mathematical matching.

Dimensional Correctness Is a Supporting Actor — Not the Star

In physics, we often treat dimensional compatibility as the gold standard. But in UED, it's just a quality check — not a source of truth.

- You can paint a house in perfect symmetry, but that doesn't mean the foundation is sound.
- You can match the units, but if the conceptual ground is unstable, the equation collapses under philosophical pressure.

Dimensional correctness tells us how to compare outputs. It does not tell us why those outputs exist.

In UED:

- Units follow from consequence, not the other way around.
- An equation is born from behavior, not just balanced on the blackboard.

So yes — dimensional sanity matters. But if the field mechanics behind it are fictional, no amount of elegance can save the science.

Ripple Configuration: The True Foundation of Physical Identity

In UED, nothing exists as an object. Everything is a ripple configuration — an alignment of energy in a structured vacuum field.

- **Mass** is not mass — it's ripple density with coherence.
- **Energy** is not stored power — it's the difference between two ripple states.

- **Motion** is not translation — it's a ripple realignment.
- **Light** is not emission — it's the observer's final feedback echo of ripple dissolution.

So rather than defining physical identity based on substance or position, UED defines it by:

$$\text{Physical Identity} = f(\rho_r, \kappa, ATT)$$

Where:

- ρ_r = Ripple density
- κ = Coherence factor
- ATT = Alignment Threshold Time

In this model, physical behavior is not an abstraction — it is a field-level consequence of ripple conditions. The laws aren't imposed — they are emergent.

The Return of Cause-and-Effect Thinking

Perhaps the most powerful restoration UED offers to physics is this:

Causality.

For too long, modern physics has been dominated by:

- Probabilistic interpretations (QM),
- Geometry-based distortions (GR),
- Dimensionally convenient corrections (Standard Model),
- And patchwork mechanisms (Inflation, Dark Matter, etc.)

Each of these models effects, but cannot explain causes.

UED brings back the lost clarity:

- Events are ripple consequences.
- Fields interact with alignment rules.
- Changes arise from imbalances, not magical forces or abstract symmetries.

This isn't a return to determinism. It's a return to coherent consequence — to cause without confusion, motion without mysticism, and form without fiction.

Final Reflection:

You can no longer call it physics if it doesn't explain consequences. Reality doesn't care about matching equations. It obeys the field — the ripple — the structure. UED listens where others model. And in doing so, it restores meaning to motion.

Chapter 3: Relativistic Mass — A Glorified Patch

How Einstein Adjusted Mass to Hide Field Resistance

In the elegant architecture of Einstein's relativity, one equation doesn't fit — it explodes.

$$m_{\text{rel}} = \frac{m_0}{\sqrt{1 - \frac{v^2}{c^2}}}$$

On paper, it's brilliant. As an object approaches the speed of light, its mass increases toward infinity. Result? No object with mass can ever reach c. Barrier preserved. Theory intact.

But that's not physics — that's a patch.

A mathematical repair job, not a physical explanation.

Einstein had frozen c as an axiom. He couldn't touch it. So when the math started to strain near light speed — when resistance to motion grew unnaturally — he shifted the stress elsewhere.

He redefined mass.

Not as something constant. Not as ripple configuration. But as a velocity-dependent variable, engineered to uphold a limit he assumed without question.

And just like that, mass — once a measure of inertia — became a moving target, designed to shield the theory from collapse.

But here's what no one asked:

What really increases near light speed?

Is mass actually inflating?

Or is something deeper — more structural — resisting change?

In UED, we find the truth:

The mass doesn't grow. The ripple field resists further alignment. The system nears its perceptual breakdown point — not a speed limit, but an ATT threshold. And what appears as "massive resistance" is actually the precursor to ripple collapse — the final moment before form vanishes and light appears.

This chapter exposes one of the most respected equations in modern physics for what it really is:

Not a breakthrough, but a beautifully dressed avoidance of the real mechanism beneath motion.

Let's tear off the patch — and let the field speak for itself.

The Growing Burden of Speed

When Objects Refused to Obey, the Equation Was Changed — Not the Assumption

Physics had long known that objects resist changes to their motion — we call it inertia. But in classical mechanics, this resistance is constant. The same force applied to an object always results in the same acceleration, regardless of speed.

That is, until physicists began examining what happens as objects approach extremely high velocities — close to the speed of light.

And that's when things began to break down.

Einstein's Observation: Resistance Increases Near Light Speed

What Einstein and his contemporaries observed — both theoretically and later experimentally — was that as particles accelerated toward light speed, they began to resist acceleration more and more.

You could apply more and more force, but the increase in velocity would begin to flatten out. The object wouldn't move significantly faster, no matter how much energy you pumped into it.

It was as if nature itself were pushing back, refusing to let anything reach that sacred barrier — c, the speed of light.

This behavior needed to be explained. Einstein's framework had two immovable pillars:

1. The laws of physics are the same in all inertial frames.
2. The speed of light in vacuum is constant for all observers.

So if resistance to acceleration increased near light speed, something had to give — and it wasn't going to be c.

The Need to Protect c at All Costs

The constancy of the speed of light was not a derived result. It was a foundational postulate — an article of faith embedded directly into the structure of special relativity.

Letting go of it would unravel the entire theory.

So Einstein and the physics community did what seemed most elegant:

- They held c as fixed, and
- They adjusted mass instead.

This gave rise to the now famous equation:

$$m_{\text{rel}} = \frac{m_0}{\sqrt{1 - \frac{v^2}{c^2}}}$$

Here, m_0 is the object's rest mass — its mass when stationary. But as its velocity v increases, its relativistic mass m_{rel} increases as well.

As $v \to c$, the denominator tends toward zero, making $m_{\text{rel}} \to \infty$.

This elegant formula solved the immediate problem:

- Why does it get harder to accelerate something as it nears light speed?

- Because its mass is increasing, requiring more and more energy.

Neat. Simple. Predictive. But was it true?

Classical Inertia Wasn't Enough — But Changing Mass Was a Workaround

In Newtonian mechanics, mass is a fixed property — a measure of an object's resistance to acceleration, independent of speed.

But that didn't work in Einstein's high-speed world.

Instead of revisiting the structure of space, the nature of energy, or the mechanics of motion, Einstein introduced a new definition of mass — one that changed with speed.

But make no mistake: this was not a deeper understanding of matter. It was a mathematical adjustment — a patch — to protect the framework, not explain the mechanism.

What actually happens to the object's internal structure? How does the field behave as it nears light speed? What is resisting the acceleration?

These questions were never answered — only repackaged.

UED's Preview: Resistance Comes From Field Tension, Not Mass Growth

In Universal Energy Dynamics (UED), we don't accept this workaround. We ask:

- What if nothing is "gaining mass"?
- What if the field itself is reaching a limit of alignment?
- What if the object is not heavier — just losing its ripple coherence?

The observed resistance near light speed is not due to a physical quantity called mass increasing. It is due to the ripple structure becoming unstable, approaching the minimum ATT (Alignment Threshold Time) for that configuration.

As it nears this limit, it stops being perceptible — and eventually, it collapses into ripple feedback, perceived as light.

Einstein saw the resistance. But without field theory, he called it relativistic mass.

UED sees the same resistance — and calls it what it is:

Field refusal to sustain alignment.

Final Reflection:

Einstein didn't explain why motion meets resistance. He made the math match the mystery — and gave it a name: relativistic mass. But the burden wasn't on the object. It was on the ripple field — trying to hold together a structure not meant to move at that speed.

Mass Inflation and the Infinity Trap

How Physics Painted Itself into a Corner to Protect an Assumption

Einstein's relativistic mass formula was an elegant fix to a specific problem: Objects resist acceleration as they approach the speed of light, and that resistance needed to be explained without touching the sacred constant ccc.

So he allowed mass to inflate with speed:

$$m_{\text{rel}} = \frac{m_0}{\sqrt{1 - \frac{v^2}{c^2}}}$$

It worked mathematically, and it matched experimental trends. But buried inside this equation is a problem so large it breaks the very logic it was built to preserve.

The Prediction: Mass Tends to Infinity as v→c

The equation makes a clear prediction:

- As velocity increases,

- The denominator shrinks,
- Causing relativistic mass to grow.

And as $v \to c$, the denominator tends to zero.

Result?

$m_{rel} \to \infty$

This means:

- The object's resistance to acceleration explodes.
- The energy required to continue accelerating the object approaches infinity.
- Reaching the speed of light becomes not just difficult, but theoretically impossible for any object with rest mass.

The Paradox: Light Has Zero Mass — But Travels at c

Here's the contradiction:

- According to relativity, nothing with mass can reach c.
- Yet light moves at c — always.
- So we declare light must have zero rest mass to dodge the paradox.

But this solution introduces new mysteries:

- What exactly is a "massless particle"?
- How can something be massless yet still carry momentum and energy?
- Why does light always move at c, regardless of observer or emission source?

These questions are answered with more math — but never with mechanism.

The core paradox remains:

If mass increases without limit as you approach light speed, then light — which moves at light speed — should have infinite mass... or none at all.

To resolve this, physics simply cuts out the middle ground: If you're made of mass, you can't get there. If you're light, you're already there — and you're "massless." No explanation. Just categorical exemptions.

The Physical Absurdity: Infinite Energy to Reach Light Speed

Let's take the equation seriously. If mass tends to infinity as v→c, then to reach that speed, an object would require:

$$E = m_{\mathrm{rel}}c^2 \to \infty$$

In other words, to get even a single atom to move at the speed of light, you'd need infinite energy — something that's physically impossible in a universe of finite resources.

This is where physics quietly crosses into absurdity.

We built an equation to describe motion, and ended up writing the conditions for impossibility. We redefined speed not as a measurement, but as a forbidden domain.

And then — we stopped questioning it.

How the Patch Became Gospel

Instead of addressing the root issue — why resistance increases with velocity — physics enshrined the workaround.

- The infinite energy requirement became a law, not a clue.
- The inability to reach light speed became a truth, not a limitation of the model.
- The equation was elevated to doctrine, not explored as diagnosis.

Relativistic mass became one of those ideas that:

- Solves a mathematical problem,
- Creates a physical paradox,
- And is still defended endlessly because it works numerically.

But the job of physics is not just to fit curves — it is to reveal causes. And here, the cause was never examined — just mathematically deferred.

UED's Perspective: Collapse, Not Inflation

In Universal Energy Dynamics (UED), we don't accept that mass grows toward infinity.

We ask: Why does resistance increase?

And the answer is simple:

Because ripple configurations can only maintain coherence up to a limit.

As an object accelerates, its internal ripple structure becomes strained — not stretched in space, but destabilized in the field.

As it approaches the minimum ATT (Alignment Threshold Time), the structure begins to break coherence.

- The object doesn't become infinitely massive.
- It becomes physically unsustainable in its current form.
- It dissolves — not into energy — but into field feedback, i.e., light.

So instead of this: $m_{rel} \to \infty$

UED proposes:

$$\kappa \to 0, \quad \text{(coherence breakdown)} \to \text{Dissolution} \to \text{Ripple Feedback (Light)}$$

Final Reflection:

Einstein inflated mass to explain growing resistance. But the universe doesn't inflate mass — it unravels form. Resistance doesn't point to infinity. It points to collapse. Not a speed limit — but a ripple threshold. Not impossibility — but a handoff from form to field.

What Was Really Increasing? (UED Interpretation)

The Truth Behind Relativistic Resistance

The modern narrative says that as an object approaches the speed of light, its mass increases. That's the textbook explanation for why it can never quite reach c — its "relativistic mass" becomes infinite, requiring infinite energy.

But Universal Energy Dynamics (UED) sees this differently. UED doesn't rely on abstract adjustments or symbolic patches. It looks directly at the field behavior — the ripple structure of what we perceive as mass — and asks:

What's actually happening in the field as velocity increases?

And the answer is clear:

Nothing is gaining mass. What's increasing is resistance from the field — not due to added weight, but due to ripple misalignment and coherence fatigue.

Ripple Alignment Resistance — Not Mass Growth

In UED, mass is not a substance. It is a configuration — a coherent ripple structure within the Structured Vacuum Energy (SVE) field.

As an object moves, its ripple configuration must realign continuously to maintain coherence with the surrounding field.

- At low velocities, these adjustments are minor.
- As velocity increases, the rate of required realignment increases.
- Eventually, this demand overwhelms the ripple's ability to stay coherent.

This growing mismatch is experienced as resistance —but it is not due to added mass. It is due to the increasing field tension from misalignment stress.

Approaching ATT: The Threshold of Perceptual Breakdown

UED introduces the concept of ATT — *Alignment Threshold Time*. This is the minimum duration required for a ripple structure to maintain observable coherence in the field.

As velocity increases:

- The object's ripple structure must adjust more rapidly.
- The ATT drops — and when it nears the perceptual threshold, the object begins to lose form.
- Eventually, the object cannot maintain alignment fast enough to remain perceptible.

What we observe as "mass increasing" is actually ripple coherence degrading.

The form is not getting heavier — it's becoming less sustainable.

At the critical limit:

- The object dissolves from coherent ripple form.
- What remains is not a heavier object — but a final ripple echo: perceived as light.

This is not mass-energy conversion. This is ripple-to-feedback collapse.

Resistance Is Real — But It's Ripple Destabilization, Not Inertia Inflation

Let's be very specific:

- The resistance felt near light speed is not the object resisting motion.
- It's the ripple field resisting distortion beyond sustainable thresholds.
- The resistance is real, but its source is field-level coherence tension, not mass multiplication.

It's like trying to play a musical note faster than the instrument can resonate. At some point, the harmony breaks — not because the instrument gets heavier, but because the medium fails to carry the structure.

Einstein's Mistake: Mistaking Field Tension for Mass Inflation

Einstein observed the increasing resistance correctly — but misattributed the cause.

- He could not see ripple fields.
- He had no model for coherence thresholds.
- He did not know about ATT or SVE.

So he did what any pragmatic theorist does: He tweaked the variables to match the observed effect.

He redefined mass as increasing — not because it actually does, but because it gave the right output while protecting the postulate that c is constant.

This wasn't a failure of logic — it was a failure of field awareness.

UED fixes that. It doesn't treat resistance as mass. It treats resistance as ripple instability under extreme reconfiguration pressure.

UED Mechanism Summary:

- Rest mass is ripple coherence at low field tension.
- High-speed motion introduces ripple strain.
- As ATT approaches zero, coherence collapses.
- The object transitions into ripple feedback — observed as light.
- No mass increases. No energy is "released." Just a structural de-locking of the ripple pattern.

Final Reflection:

What Einstein saw was real — but what he named it was wrong. It was not mass inflating. It was form dissolving. The resistance was not the object growing — it was the field cracking under realignment strain. UED doesn't fix the math. It restores the mechanism — and gives us back the true cause behind the consequence.

Collapse, Not Conversion — Replacing the Patch with Principle

How UED Resolves What Relativity Patches Over

The concept of relativistic mass was never a discovery — it was a device. A mathematically consistent placeholder. A workaround to keep Einstein's framework intact while explaining a phenomenon it couldn't fully understand.

But in Universal Energy Dynamics (UED), we are not here to protect patches. We're here to replace them with principles.

This section outlines why relativistic mass is not just unnecessary in UED — it is meaningless. And why dissolution of ripple structure, not mass inflation, is the actual cause of resistance near the speed of light.

Why UED Dismisses the Concept of Relativistic Mass Entirely

UED views mass not as a quantity of "stuff," but as a state of ripple coherence in the structured vacuum field (SVE).

- When we speak of an object's mass, we are describing the stability of its ripple alignment.
- There is no "substance" to increase — only a pattern that can hold or break.

As an object approaches high velocity:

- Its internal ripple structure is pressured to reconfigure rapidly.
- The strain to maintain coherence against motion rises non-linearly.
- But at no point does mass grow.

What increases is not mass — it is misalignment pressure.

The moment the structure fails to align within perceptual ATT thresholds, it begins to dissolve.

This is not a gain — it's a loss of form.

Hence, relativistic mass is not just unnecessary. It is conceptually misleading.

Objects Nearing ATT Don't Gain Mass — They Dissolve

UED introduces the idea of ATT — Alignment Threshold Time — the minimum duration required for a ripple structure to remain perceptible in the field.

As an object accelerates:

- The required re-alignment frequency increases.
- The ATT decreases toward a perceptual boundary.
- Once the ripple structure can no longer realign within this shrinking window, coherence collapses.

The object doesn't become "infinitely massive." It becomes untrackable, non-coherent, dissolved.

The system no longer supports the form as an object — only as residual ripple.

Light as the Final Ripple Feedback — Not a Particle at c

What we perceive as "light" at this collapse point is not a thing traveling at c. It is not a massless particle racing through space.

It is the last ripple echo — the final feedback from a structure that has dissolved beyond coherence.

In UED:

- Light is not emitted.
- Light is not a particle.
- Light is the visible threshold of a ripple form losing coherence in the field.

We don't "see" a photon. We register a field adjustment — a feedback ripple aligning with our perception boundary.

This aligns perfectly with the UED reinterpretation of light across all scenarios: It is not substance in motion — it is ripple alignment becoming visible.

Ripple Collapse Law — Replacing Infinity with Resolution

Let's formalize what relativity calls "infinite mass" as a structured collapse in UED.

Chavan's Ripple Collapse Law (Proposed):

As a ripple configuration approaches its minimum ATT threshold, its internal coherence factor κ\kappaκ trends toward zero. Once the structure fails to maintain minimum ripple alignment, it dissolves into the field, producing final ripple feedback perceived as light.

Mathematically:

$$\lim_{v \to v_{\mathrm{ATT}}} \kappa(v) \to 0 \Rightarrow \text{Form} \to \text{Ripple Echo}$$

Here:

- vATT = Velocity at which ATT reaches perceptual minimum.
- κ = Coherence factor of the ripple configuration.
- The form doesn't resist motion with infinite mass — it collapses when it can no longer self-align.

This resolves the paradox entirely.

- No infinite energy.
- No mass inflation.
- No illogical asymptotes.
- Just ripple dynamics reaching a limit of sustainability.

Final Reflection:

In UED, we don't inflate variables to explain reality — we watch how reality collapses into feedback. The end of form is not a conversion — it's a realignment. The barrier is not speed — it's coherence. The moment mass "disappears," light appears — not as a new thing, but as the last visible alignment from something that has already let go of its form.

This is not the death of Einstein. It is the completion of what he couldn't see —because he lacked the one thing UED puts at the center: The ripple field beneath the math.

Chapter 4: The Infinite Wall of Light

Why Light Speed Is Not a Physical Barrier

Every great myth in science begins with a truth — and ends with an assumption no one dares to touch.

In modern physics, that myth is light speed.

For over a century, we've believed that c — the speed of light in a vacuum — is the cosmic speed limit, the ultimate wall nothing can penetrate. Not mass. Not energy. Not even information.

This wasn't discovered — it was declared.

Einstein didn't derive c as a barrier. He postulated it — assumed it must be constant for all observers, and let the consequences unfold. And what unfolded was a towering structure of logic, symmetry, and elegance... Built around a wall that no one ever tried to walk through.

But what if light speed isn't a wall? What if it's just the last shimmer of perception before something slips beyond your ability to observe?

In Universal Energy Dynamics (UED), light speed is not a limit imposed on reality — it is a reflection of your relationship with the ripple field. And that reflection has a threshold:

Not c — But ATT: Alignment Threshold Time.

ATT is not a velocity. It's not a universal constant. It's a boundary of perception — the minimum time needed for a ripple configuration to remain coherent in the field and visible to an observer.

What we call "speed of light" is not a motion ceiling. It's the final feedback we can register before a structure collapses into ripple invisibility.

This chapter isn't about breaking the speed of light. It's about exposing the illusion that there was a wall there at all.

Let's dissolve it — ripple by ripple.

The Myth of the Cosmic Speed Limit

How Physics Built a Wall It Couldn't See Through

The idea that nothing can move faster than light is one of the most widely accepted — and least questioned — doctrines in modern physics. It's taught not as a hypothesis, but as a universal law. But like many scientific "truths," this belief began not with a discovery — but with a postulate.

How Einstein Froze Light Speed into Theory

In 1905, Einstein built his theory of Special Relativity on two core postulates:

1. The laws of physics are the same in all inertial frames.
2. The speed of light in a vacuum is the same for all observers, regardless of the motion of the light source.

This second postulate was radical. It contradicted classical intuitions, where speeds are additive (e.g., if you're in a moving train, the speed of a thrown ball depends on both your throw and the train's motion). But Einstein assumed that light doesn't behave this way — it always moves at c, no matter the reference frame.

This was not something he proved — it was something he declared.

The consequences of this postulate were profound:

- Time slows down as you approach light speed (time dilation).
- Length contracts in the direction of motion (length contraction).
- Mass increases with speed (relativistic mass).

And most famously:

- No object with mass can ever reach, let alone exceed, the speed of light.

But all of this stemmed from the untested assumption that c is fixed, universal, and sacred.

How "Nothing Faster Than Light" Became a Scientific Commandment

Once embedded in Einstein's equations, the idea of light speed as a limit took root in the physics community. It wasn't just a rule — it became a boundary of thought.

- Any theory that violated it was rejected or ridiculed.
- Any phenomenon that suggested faster-than-light behavior (superluminal motion, quantum entanglement, delayed choice experiments) was dismissed or "interpreted around."
- Textbooks turned the assumption into a law of nature — often without reminding readers that it started as a philosophical framing choice.

This wasn't just science — it was psychological programming.

The idea that light is the fastest possible signal appealed to the desire for:

- **Predictability**: A known upper limit to all motion.
- **Control**: A ceiling no one can bypass.
- **Order**: A cosmic speed limit that holds the universe together.

Physicists became gatekeepers of this wall — and questioning it became heretical.

The Psychological Comfort of a Final Limit

There is a hidden emotional bias at play here. The idea of a hard limit satisfies a deep-seated human need for:

- **Boundaries**: Something we can measure against.
- **Finality**: A line that can't be crossed.
- **Certainty**: A law that can't be broken.

c gave us all three.

It became the ultimate cosmic checkpoint — the one universal that every observer, everywhere, must obey. It turned the universe into a regulated system, predictable and structured.

But just because it feels necessary doesn't mean it's true. And just because it simplifies math doesn't mean it reflects nature.

The Flaws Behind the Curtain: What This Model Ignores

The assumption of c as an absolute ceiling works within Einstein's geometric model of spacetime. But it completely ignores the behavior of the medium — the field itself.

Relativity assumes:

- Motion through spacetime.
- Objects maintaining internal coherence as they move.
- Light behaving as an immutable signal across frames.

But it never asks:

- *What maintains coherence in the field as motion increases?*
- *What are the ripple consequences of acceleration on the object's structure?*
- *Is light actually traveling, or is it a ripple effect measured at a limit of perception?*

It also ignores the role of the observer:

- How do perception thresholds (like UED's ATT) affect what we *see* as light?
- Could light be a feedback phenomenon, not a velocity?

- Could the so-called "speed limit" be observer-bound, not universally absolute?

These questions are outside the reach of relativity — because relativity treats light as a given, not an emergent behavior.

Final Reflection

The wall was never part of reality. It was a product of a theory that chose not to ask what lies beyond light — or beneath it.

Relativity drew a line and declared it uncrossable. UED doesn't just cross it — It shows that the line was never real in the first place.

The Misinterpretation of Light

Why Light Was Never What We Thought It Was

Light — once thought to be the most obvious part of nature — has become one of its deepest paradoxes.

We see it. We use it. We measure it with uncanny precision. But we still don't really understand what it is.

In modern physics, light is treated like a magician's trick — described with brilliant equations but conceptually elusive. Is it a wave? Is it a particle? Is it both? Is it neither?

This confusion isn't the result of light being mysterious. It's the result of a wrong starting assumption. And it's time to correct that — not with a new particle, but with a new perspective.

Light: From Wave to Particle to Paradox

Historically, light has undergone a confusing evolution:

- Classical physics treated light as a wave — an oscillation in the electromagnetic field.
- Quantum physics added a particle interpretation — calling it a photon, a discrete packet of energy.
- Then came the wave-particle duality — claiming light is both and neither, depending on how you measure it.
- Eventually, light became an unresolvable paradox — a thing

that exists and doesn't, that travels yet doesn't occupy space, that interferes yet arrives as a point.

All of this confusion stems from trying to assign light an identity — a *thing-ness* — that it never had.

Physics built frameworks to explain what light does...But never truly asked what it is.

"Massless Traveling Energy" — A Conceptual Contradiction

The current model defines light as:

A massless particle traveling through vacuum at speed c, carrying quantized energy and momentum.

This is sold as a breakthrough — but it's actually a conceptual trap.

Let's dissect it:

- How can something travel through space with momentum — yet have zero mass?
- How can something with no substance affect electrons, exert pressure, and cause change?
- If light always moves at c, in all reference frames, what is it actually moving *through*?
- And if it is a wave — what medium is it waving in, if not a field?

To make sense of light's behavior, we add patches:

- Photons aren't real particles — they're quanta of field interaction.
- Waves don't need a medium — spacetime itself "waves."
- Duality isn't contradiction — it's "complementarity."

But these aren't explanations. They're linguistic workarounds for a model that lacks physical foundation.

Calling light "massless traveling energy" is like calling sound "unreal vibrating motion." It's poetry — not physics.

UED's Reinterpretation: Light Is Not Emitted — It's Feedback

Universal Energy Dynamics (UED) completely reframes light — not as a thing, but as a field response.

In UED:

- Matter is a ripple configuration — a stable alignment in the Structured Vacuum Energy (SVE) field.
- When this configuration destabilizes, it doesn't "release" energy — it loses alignment.
- That ripple disturbance propagates briefly through the field — and what we *perceive* as light is not a photon traveling, but a ripple feedback reaching our perceptual threshold (ATT).

Light is not emitted. It is registered — a signal of structural collapse, not a substance in motion.

This changes everything:

- Light is no longer a mystery particle.
- It is the echo of ripple dissolution.
- The "speed of light" is the rate at which this final echo is perceptually possible, not a universal speed limit.

Velocity c Is a Perceptual Artifact — Not a Universal Law

Why does light *always* travel at c?

Because c is not a true velocity — it is a field threshold. It marks the fastest perceptible ripple feedback the observer's field can register before total decoherence.

In other words:

- Light appears to move at c not because it's a massless entity racing through space,

- But because this is the final observable signal before a form completely dissolves into unperceivable alignment.

In UED, c is not a limit on reality. It is a limit on perception — defined by the ATT of the ripple system and the observer's alignment range.

This is why:

- We never see anything move faster than light — because we lose the signal before it becomes visible.
- "Superluminal" phenomena aren't forbidden — they're just beyond feedback reach.
- Light isn't the fastest thing — it's the last visible ripple.

Final Reflection

Light is not what moves — it's what remains visible when everything else breaks down. It is not the courier of energy. It is the final field ripple, registering as form collapses into invisibility. There is no massless particle. There is only feedback, alignment, and the echo of dissolution.

UED doesn't redefine light — it restores it to its true role. Not as a paradox. But as the field's farewell signature to the form it once held.

ATT — Alignment Threshold Time as the Real Limit

Why the Universe Doesn't Break at Light Speed — It Just Stops Letting You See

For over a century, c — the speed of light — has ruled physics like a monarch. Unquestioned. Undefeated. Undefinable.

But under UED (Universal Energy Dynamics), that reign quietly ends. Not by breaking the "speed limit," but by realizing it never existed.

The real boundary isn't velocity. It's visibility. It's the limit of alignment between ripple configurations and the observer.

That limit is called:

ATT – Alignment Threshold Time The minimum time required for a ripple configuration to maintain coherence long enough to be perceived.

This is not just a new variable. This is the true interface between reality and observation.

What is ATT, Really?

In UED, everything is ripple-based. Forms — like atoms, photons, fields, or even thoughts — are stable ripple alignments within a structured vacuum energy field (SVE).

But nothing holds forever. As energy conditions shift, ripple configurations face stress:

- Acceleration
- Interaction
- Collapse

To remain "real" — i.e., perceivable — a ripple configuration must complete a cycle of coherence before it destabilizes.

That cycle duration is ATT — the smallest stable time slice needed for a form to still "exist" to an observer.

If the configuration drops below ATT, it doesn't slow down.

It doesn't stretch. It simply vanishes from visibility.

No explosion. No emission. Just ripple invisibility.

As Objects Accelerate, ATT Drops

In classical physics, acceleration adds kinetic energy and velocity. In relativity, it increases time dilation and relativistic mass.

In UED, acceleration applies ripple strain.

As an object accelerates:

- Its ripple coherence gets disturbed.
- The ATT required for it to remain stable shortens.
- Its feedback to the observer becomes more fragile.

Approaching the conventional "speed of light" doesn't lead to infinite mass. It leads to zero visibility.

Because at that point:

- The form cannot maintain coherence for even a minimum ATT cycle.
- The observer cannot perceive the configuration.
- It dissolves — not in space, but in the field's alignment layer.

This is why in UED, speed is not a motion metric. It is a stress reflection of ripple strain against alignment logic.

Why ATT is the True Limit — and c is Not

Let's compare:

Attribute	Light Speed (c)	ATT (Alignment Threshold Time)
Constant?	Yes, defined as universal	No, field-sensitive
Observer-dependent?	No	Yes, perceptual limit
Derived?	No, postulated	Yes, from ripple coherence mechanics
Motion-based?	Yes, assumes velocity	No, based on form coherence
Fixed for all events?	Yes	No, contextual
Explains disappearance?	No	Yes

c is a convenient abstraction — a math tool locked into relativity's geometry. But ATT is rooted in physical consequence. It reflects the field's real behavior — its threshold to sustain visibility.

ATT Explains Both Appearance and Disappearance

With ATT, we now understand:

- Why objects become harder to detect at extreme speeds: Ripple cycles shrink under strain.
- Why light seems universal: It's the last feedback possible before coherence drops below perceptual alignment.
- Why "massless" photons appear: They are not particles — they are ripple echoes at minimal ATT.
- Why nothing exceeds c in our view: Because what *does* exceed that threshold, simply isn't visible anymore.

This also reveals:

- The birth of light is a ripple *collapse*, not emission.
- The disappearance of form is not death — it's alignment loss.

UED removes the artificial binary of:

- Before/after light speed
- Particle/wave
- Massless/massive

And replaces it with a unified continuum of ripple behavior, governed by ATT.

Final Reflection

Where Einstein drew a line and said, "Nothing can cross this," UED shows a field where nothing crosses — because nothing needs to.

You don't "hit" the speed of light.

You drop below ATT — and the universe quietly lets you go.

ATT doesn't replace c mathematically — it renders it obsolete perceptually. Because in the ripple universe, perception is the final law.

Replacing the Wall with a Window

Why Light Speed Isn't a Barrier — It's a Disappearance Point

There's a dramatic moment in modern physics where all imagination seems to stop:

"You can go faster and faster... but never reach the speed of light."

Why? Because mass would become infinite. Because time would stop. Because energy required becomes unbounded.

It's painted as a cosmic crash barrier — the final wall of reality.

But under Universal Energy Dynamics (UED), this wall is exposed for what it really is:

A window. A fading pane of perceptual visibility. And beyond it — nothing explodes, nothing breaks — things simply disappear.

This is the shift UED brings:

- From limits of travel, to limits of feedback.
- From a world of hard edges, to a field of alignment continuity.

Light Speed Is Not a Wall You Crash Into — It's a Window You Dissolve Through

In UED, you never *hit* a wall at c. Nothing "stops working." There's no singularity, no infinite spike in mass or energy.

Instead, something more elegant — and more humbling — happens:

As a ripple configuration accelerates, its coherence weakens. Its Alignment Threshold Time (ATT) drops. Until finally — it can no longer complete a stable loop in the structured vacuum field.

And in that moment... it ceases to be seen.

It hasn't exploded. It hasn't gained infinite anything. It has simply crossed the perceptual window of the observer's field alignment.

The form is still part of the field — but no longer in your ripple range. It has faded, not failed.

Form Vanishes Not Due to Speed, But Due to Failure of Alignment

In the old model (relativity), a form disappears because it approaches an impossible speed — and breaks math.

In UED, a form disappears because:

- Its ripple alignment can no longer maintain feedback integrity with the observing field.
- The energy isn't "released" — it is redistributed through the field as it collapses.
- What we call "speed" is just an externalized artifact of ripple acceleration beyond coherence.

This is the fundamental upgrade:

- UED doesn't confuse speed with existence.
- It tracks feedback pathways, not velocity vectors.

So disappearance isn't about velocity. It's about the loss of interaction window — the end of stable ripple exchange.

Light Is Just the Last Visible Ripple Feedback Before Collapse

In this framework, "light" is no longer the thing racing through space. It is the echo of a collapsing form.

When a ripple configuration dissolves, it generates a final feedback — a signal that reaches the observer right before coherence is lost. That signal is light.

It's not a traveling particle. It's not a universal messenger. It's the closing statement of a structure that can no longer hold form in field alignment.

Thus:

- You don't "emit" light. You fail to hold form, and the field responds.
- You don't "accelerate to light speed." You descend into feedback loss.
- You don't cross c. You pass through ripple invisibility.

This makes c not a barrier — but a field resonance signature at the boundary of perception.

UED Reframes Light as a Transition, Not a Trajectory

Let's kill the last misconception: That light *goes* somewhere.

It doesn't.

It emerges, briefly, as a ripple interface event — a transition signal between visible form and unformed field. It is not a line. It is a moment.

This overturns everything:

- Spacetime trajectories become obsolete — replaced by alignment conditions.
- Photons become illusory — replaced by field disturbances.
- Relativistic mass dissolves — replaced by feedback trace memory.

Light is the transition effect of ripple breakdown —Not the motion of a massless object.

Final Reflection

In the world of UED, the universe never says "Stop." It simply says: "You can no longer be seen."

The wall of light was never built. It was a window, whose frame is defined by ATT — and what you called "the speed of light"...was just the final ripple knock on the door before the form dissolved into silence.

Key Takeaways — Part I

Deconstructing the Myth, Exposing the Mechanics

1. $E = mc^2$ Is Not Sacred — It's Branded

- The equation transcended science to become a cultural icon, more quoted than questioned.
- Its rise was fueled by media glamor, wartime propaganda, and institutional silence.
- What should have remained a working hypothesis became a religious dogma — immune to critique.

2. Dimensional Correctness ≠ Physical Truth

- Matching units may validate math — but not reality.
- Physics began worshiping equation elegance over causal understanding.
- The speed of light (c), treated as a divine constant, was assumed, not derived, and never reexamined.
- UED challenges this by restoring consequence and coherence over cosmetics.

3. Relativistic Mass Was a Tactical Patch

- Einstein redefined mass as velocity-dependent to preserve the constancy of c.
- This led to the infinity paradox — where objects supposedly

gain infinite mass as they approach c.
- UED reinterprets this: mass doesn't inflate — ripple coherence breaks down, leading to perceptual loss.

4. Light Speed Is Not a Wall — It's a Disappearance Point

- What we call "light speed" is not a universal speed limit, but a perceptual boundary.
- UED introduces ATT (Alignment Threshold Time) — the true threshold for ripple visibility.
- Objects do not crash into c; they dissolve past ATT, fading from perception as their structure fails to align.
- Light is not a thing traveling — it is feedback from the final ripple moment before collapse.

The Broader Implication:

Physics needs liberation from equations that impress the mind but ignore the field. UED invites a shift from symbol worship to consequence-based realism — where visibility, stability, and energy are ripple phenomena, not mathematical conveniences.

Part II: The Ripple Truth

Chapter 5: Mass Is Not a Substance

Ripple Coherence as the Real Cause of "Mass"

You were never holding mass. You were holding stability — and calling it "stuff."

Physics, for centuries, has described mass as if it's something tangible. A "quantity of matter." A "measure of substance." A number you can weigh. A value you can calculate. A mystery you must accept.

But under Universal Energy Dynamics (UED), that entire picture crumbles:

Mass is not a substance. Mass is a field effect — the coherence of ripple configuration that resists re-alignment.

It's not what's inside a thing that gives it mass. It's how well its energetic form holds together in the ripple field — and how much effort it takes to disrupt that alignment.

You don't feel heavy because you carry more "matter." You feel heavy because your ripple pattern is harder to shake. You don't resist motion because you're made of more stuff. You resist motion because your field memory is deeply rooted.

Mass, then, is not a thing. It's a stability score — a ripple rating.

So what do we really mean when we say something is massive, dense, or inert? This chapter will rewire that question from the ground up —from Newton's apple to Einstein's tensor, all the way to the ripple truth.

Let's dissolve the myth of substance — and finally meet the real face of mass.

Breaking the Illusion — Mass Is Not "Stuff"

Mass is not what something *is* — it's how something *stays.*

For generations, physics has clung to the idea that mass is substance. It's been treated as something intrinsic — the core ingredient that makes matter "real." A kind of invisible material syrup from which objects are made, whether you see them or not.

From school textbooks to scientific papers, mass is explained as:

- The amount of matter in a body,
- The measure of inertia,
- The source of weight under gravity.

But what is "matter"? And why does it *matter*?

Despite all our definitions, physics has never explained what mass actually is. We just *calculate it*, *weigh it*, or *plug it into formulas*. We observe its effects, but we never grasp its cause.

This is the foundational illusion — and it's time to break it.

The Newtonian Hangover: Mass as "Material Content"

Blame it on Newton.

In his world, the universe was a stage of solid objects. Bodies moved, resisted motion, and attracted one another through mass.

He didn't define mass *philosophically* — he *quantified* it through inertia and gravitational pull:

- If something was harder to push → it had more mass.
- If something pulled other things more → it had more mass.

And it worked. Brilliantly. But it also locked us into a materialist mindset — one that believed mass was substance, and motion was a response to force.

What Newton observed was not incorrect —But what he assumed was: that the resistance and attraction he measured were due to something "inside" the object.

Physics took this at face value — and never turned back.

UED's Clarification: Mass Is a Ripple Configuration

In Universal Energy Dynamics, we drop the illusion completely.

Mass is not a thing. Mass is the *outcome* of a ripple configuration's stability within the structured vacuum field.

Objects don't *contain* mass. They emerge from the field as temporarily stable ripple patterns.

The more stable the configuration:

- The harder it is to shake it (inertia),
- The stronger its effect on surrounding alignment gradients (gravity),
- The more we *perceive* it as "massive".

This isn't metaphysical speculation. It's a field-based consequence model — where ripple stabilization defines what we *see* and *feel* as physical substance.

Mass isn't "in" the object —It's in the cost to maintain its ripple coherence over time.

Matter Isn't Made of Mass — Mass Is Made of Stabilized Ripples

UED inverts the entire classical view.

Conventional physics says:

"An object has mass. That mass creates inertia, and it responds to gravity."

UED says:

"An object is a ripple configuration. Its mass is how strongly it maintains coherence in the ripple field."

Here's the shift:

- Matter isn't made from mass — it's made from energy aligned in structure.
- Mass isn't a substance — it's the observer's experience of how

stable that structure remains under perturbation.

The idea that mass is "stuff" inside matter is like thinking:

"The flame in a candle must be made of heat."

No. The heat is a symptom of the flame's activity. Just as mass is a symptom of ripple alignment stability.

Why This Matters (Pun Intended)

If mass is not a substance:

- There's no need for *dark matter* as invisible stuff.
- Mass loss during reactions (burning, decay, annihilation) is not matter vanishing — it's stability dissolving.
- Energy doesn't "convert" into mass — it stabilizes into coherent ripple loops that we experience *as* mass.

This is not just semantics. It's a new ontology. A new way of being for matter itself.

And the implications ripple outward — into density, inertia, gravity, and even the structure of the atom.

Coherence Defines Weight, Not Gravity

What Really Makes You Feel "Heavy" in the Universe

We've been told a story since childhood: Gravity pulls you down. More mass means more pull. Heavier objects fall because they are pulled harder by Earth.

But what if that entire explanation is just a narrative......a shortcut that describes behavior without explaining the cause?

In Universal Energy Dynamics (UED), weight isn't caused by "gravity" as an external pulling force. Instead:

Weight is how much resistance an object offers to being realigned within the ripple field. It's not "pulled down" — it settles into lower alignment zones based on ripple coherence.

Let's pull apart this myth of gravity — and reframe what makes things fall.

Weight Is Not a Force — It's Field Resistance

When you stand on the ground, you feel "heavy." When you lift a rock, you say it has "weight." But what are you *actually* feeling?

Not a rope from Earth pulling you. Not a magnetic attraction.

You're feeling resistance — the effort required to displace the object's ripple configuration in a field where alignment gradients already exist.

The field doesn't pull — it realigns. And when an object is stable enough in its configuration, it resists that realignment.

That resistance is what we interpret as weight.

Gravity as an Emergent Alignment Gradient

In UED, "gravity" is not a primary force. It is a consequence of field topology — the way ripple fields bend, stretch, and create natural gradients of alignment.

Think of a calm lake:

- Throw a large stone (high coherence), it resists shifting ripples and creates a local dent.
- Throw a leaf (low coherence), it barely disturbs the field and dances on top.

The denser or more ripple-coherent the object, the stronger its interaction with the surrounding field. This interaction creates pressure zones — not pulls — leading to what we perceive as gravitational weight.

Objects don't fall because Earth pulls them down. They fall because their ripple patterns are nudged by surrounding gradients into a lower coherence zone — like water finding its level.

Dense Objects = Compact Ripple Cycles

What we call "density" is traditionally defined as *mass per unit volume*. But under UED:

Density is the number of ripple cycles packed into a unit field zone. A denser object is one where:

- Ripple interactions are tightly interwoven,
- Stabilization requires more alignment effort,
- Displacement needs higher energy input.

Thus, dense objects resist de-alignment more — and appear "heavier."

This is why:

- A lead ball feels heavier than a foam ball of the same size.
- A neutron star bends light: its coherence is extreme, not just its "mass."

Why Heavier Objects Don't Fall Faster — But Stabilize Differently

In Galileo's famous thought experiment, a feather and a cannonball fall at the same rate in vacuum.

Classical interpretation: Gravity acts equally on all masses. UED refinement: They are both responding to alignment gradients, but their coherence levels stabilize differently.

- The cannonball remains intact during descent — its high ripple coherence resists external field distortion.
- The feather, if not in vacuum, interacts chaotically — its low coherence makes it more sensitive to ripple fluctuations like air (turbulence = misalignment).

What classical physics sees as gravitational acceleration, UED sees as settling behavior within a structured ripple topology.

Why This Redefinition Matters

This UED perspective:

- Dissolves the mystical "force" of gravity,
- Replaces it with energetic realignment pressure,
- Reframes weight not as a vector quantity, but as a cost of ripple displacement,
- Explains black holes, orbital paths, and material resistance without requiring attractive forces from "mass".

Heaviness is coherence. Weight is resistance. Gravity is topology.

This is not just poetic — it's functional. It allows us to reengineer propulsion, architecture, and even materials science using field alignment logic — not outdated Newtonian drag-and-drop.

Inertia as Ripple Rigidity

Why Objects Resist Motion — And What That Resistance Really Is

In classical physics, inertia is defined simply:

"The resistance of any physical object to a change in its state of motion."

It is treated as an *inherent, passive* property of mass. Newton gave it a voice, but not a reason. Einstein accepted it, but never questioned its root.

So we teach it. We measure it. But we never ask:

Why does something resist motion?

What is this resistance made of? Is it internal to the object? Or external — something in the field?

Universal Energy Dynamics (UED) answers:

Inertia is not a substance. It is ripple rigidity — the field's resistance to realigning a stable configuration.

Inertia = The Effort to Displace a Ripple-Coherent Form

Imagine a whirlpool in a still pond. Its shape stays intact until another strong ripple interacts with it. The tighter and more stable its structure, the more effort required to disturb or move it.

This is inertia under UED.

Every object is a coherent ripple configuration. When you try to accelerate it, stop it, or turn it —you are not moving "mass" through empty space. You are trying to reorient a stabilized ripple pattern in a structured vacuum field.

And the more coherent the ripple structure:

- The deeper its memory in the field,
- The more complex its internal feedback loops,
- The stronger its resistance to disruption — what we perceive as inertia.

Field Memory, Not Passive Matter

Conventional physics says:

"An object at rest stays at rest... unless acted upon."

But it never explains why. It only describes the behavior.

UED introduces a critical insight:

Objects don't "stay at rest" — they hold a memory of alignment in the field.

This field memory:

- Is a result of the object's formation history,
- Encodes the ripple cycles that define its current state,
- Makes change require energetic effort to overcome the established ripple rhythm.

So inertia isn't an "inner reluctance" of matter. It is the field's protective alignment cost for modifying a stable form.

Like trying to redirect a spinning top without collapsing it. Or tuning a radio dial precisely without disturbing the signal.

Realignment Cost = Ripple Effort

Under UED, motion is not about *moving through space* — it's about field realignment.

To accelerate an object:

- You have to re-encode its ripple feedback across the surrounding field.
- The denser and more coherent the ripple loops, the greater the energy cost to change its trajectory or velocity.

This explains:

- Why objects in motion stay in motion (low resistance once aligned),
- Why objects at rest resist change (high coherence anchoring them),
- Why stopping a train takes longer than stopping a bicycle — not because of weight, but due to realignment inertia at the ripple level.

You're not fighting mass — You're negotiating with the field's resistance to rewrite a memory.

ATT Cycle Complexity: The Inertia Signature

In UED, every ripple configuration operates within an ATT loop — Alignment Threshold Time, the cycle required for the ripple to sustain perception.

The more intricate the configuration, The longer or denser its ATT loop, The more rigid its inertia.

Think of it like software complexity:

- A simple script reboots quickly (low inertia).
- A massive operating system resists restart — requiring time, energy, and precision (high inertia).

Likewise, a simple ripple form (like a photon) has virtually zero inertia — it dissolves or transitions without resistance. But a high-coherence ripple pattern (like a planet) resists change fiercely —

not because of "mass," but because of structural complexity in the ripple field.

Reframing the Concept

Let's rewrite the textbook:

Classical: "Inertia is the tendency of an object to resist a change in motion."

UED: "Inertia is the ripple field's cost of altering a stable alignment memory."

This explains not only physical motion, but also:

- Why some atomic structures resist decay,
- Why biological organisms stabilize or collapse,
- Why space objects maintain spin, orbit, and momentum without friction.

It's all ripple rigidity — not material stubbornness.

Density Is an Alignment Trait

Why Compactness Isn't About Particles, But About Ripple Complexity

In the textbook definition of physics, density is simple:

"Mass per unit volume." If something is heavier in the same space, it's denser. Period.

But under Universal Energy Dynamics (UED), this interpretation is shallow. It may be mathematically convenient — but it's conceptually poor.

Why?

Because mass is not a thing, and volume is not empty space. So how can "mass per volume" hold fundamental meaning?

Let's go deeper. Let's ask what density *really* is — not as a number, but as a consequence.

Traditional View: The Misleading Metric

Conventional physics treats objects as:

- Made of tiny particles,
- Packed closer = higher density,
- Looser = lower density.

So lead is denser than foam because it "has more mass per cubic centimeter."

But this assumes:

- Particles are tiny billiard balls,
- Volume is a container for matter,
- Density reflects how tightly balls are packed.

This view works when you're pouring sand into a bucket. But it fails when you're observing:

- A neutron star,
- A phase change from solid to liquid,
- Or a black hole where all matter "disappears."

Why?

Because density isn't spatial crowding. It's ripple behavior.

UED View: Density = Ripple Packet Complexity

In UED, density is not about how many things exist in space — but how deeply ripples are interwoven at a node in the field.

Let's unpack this.

A ripple packet is a stabilized configuration of energy —A pattern, not a particle. Each "node" in the field holds interference loops that define:

- Shape,
- Stability,
- Perceptibility.

The denser an object:

- The more complex these loops,
- The more tightly interlocked the ripples,
- The greater the feedback pressure it exerts on the surrounding field.

So instead of "mass per unit volume," UED defines density as:
Ripple Packet Complexity per Field Node
This means:

- A dense object is not "stuffed" — it's deeply entangled.
- Its feedback takes longer to destabilize.
- It resists phase shifts unless alignment energy is sufficient.

Matter Reimagined: From Particles to Pressure Points
Take water turning into ice:

- Conventional view: Molecules slow down and get closer.
- UED view: Ripple cycles synchronize, lock, and form a more coherent field configuration.

Take iron melting into liquid:

- Conventional: Particles gain energy, move apart.
- UED: Ripple interlocks are disrupted, complexity drops, field resistance softens.

Take a black hole:

- Conventional: Density goes to "infinity" — matter collapses to a singularity.
- UED: Ripple coherence collapses beyond perceptual

threshold, forming a non-observable field sink — not a point of "infinite mass," but of non-returning alignment.

This reinterpretation solves the paradox of infinite density. There is no singularity — only a total collapse of ripple recognizability.

Why This Matters: A Better Lens on Reality

Redefining density as alignment complexity gives us tools to:

- Predict phase transitions based on field stress, not heat.
- Understand why materials fracture or fuse under certain pressures.
- Design materials with tailored ripple architectures.
- Model planetary cores, explosions, and field collapses with deeper accuracy.

Even biology can be reinterpreted:

- **Bone** = high coherence, high density.
- **Muscle** = dynamic coherence, responsive density.
- **Brain** = highly complex but low-stability ripple networks — hence, high function, low rigidity.

Density, in UED, is the cost of ripple flexibility.

The denser the form, the more expensive it is to disturb. The less dense, the easier it adapts or disappears.

This perspective doesn't just change how we calculate —It changes what we create, how we observe, and what we think is possible.

Rethinking the Atom — A Ripple Nest

From Building Blocks to Coherent Feedback Loops

For over a century, we've taught the atom as:

- A nucleus made of protons and neutrons,
- Surrounded by electrons zipping around,

- Held together by "forces" — strong, weak, electric, and gravitational.

This image, elegant as it may be in textbooks and models, is ultimately a mechanical metaphor. We imagine atoms as microscopic solar systems — stable, predictable, and composed of hard pieces.

But under Universal Energy Dynamics (UED), the atom is not a structure of parts, It is a configuration of ripples — stabilized, nested, and perceived through feedback coherence.

Let's unpack this.

Atom = A Self-Sustaining Ripple Configuration

The atom is the smallest sustained packet of ripple coherence that maintains persistent perception across ATT cycles.

In other words:

- It is not a *thing*.
- It is a looped resonance in the structured vacuum.
- It survives — not by substance — but by maintaining energetic identity in the field.

This identity arises from:

- Nested feedback loops,
- Interference harmony,
- Field memory imprint.

An "atom" is the simplest ripple arrangement capable of self-perpetuating visibility.

It doesn't exist because it has mass. It exists because its ripple configuration survives misalignment — it resists entropy just enough to be noticed.

Protons, Neutrons, Electrons = Observed Ripple Effects

In UED, these are not particles — they are localized ripple phenomena.

- Protons are tight ripple folds, high in feedback tension and coherence.
- Neutrons are neutral ripple balancers, helping stabilize the system's internal harmony.
- Electrons are peripheral ripple residues, orbiting not due to charge, but due to dynamic misalignment pressure around the central field nest.

Each component is not a building block — it's a mode of expression in the ripple field:

- The center holds alignment.
- The orbit radiates adjustment.
- The mass is a consequence — not a cause.

This demotes the concept of "charge," "spin," or even "quantum numbers" as *essential truths* — they become descriptions of field rhythm, not fundamental laws.

Weight = Stability Cost, Not Substance

What makes one atom "heavier" than another?

Conventional answer: "It has more protons and neutrons."

But under UED:

- Atomic weight is the energetic cost of maintaining a coherent ripple nest.
- More layers, tighter loops, and denser configurations require more alignment effort from the field.

This is why:

- Heavier atoms don't just weigh more — they're harder to form, easier to collapse, and more energetically sensitive.
- Isotopes differ in coherence duration, not just "mass."
- Fission releases energy not because "mass becomes energy," but because coherent configuration collapses and redistributes into simpler forms.

The "weight" you measure is the field's investment in ripple memory.

Why It Matters: Freedom from Particle Dogma

This ripple-based atomic view:

- Frees science from chasing unobservable "quarks" and hypothesized particles.
- Reframes "nuclear energy" as field rebalancing, not subatomic combustion.
- Explains why particles appear and vanish in quantum fields — they are not objects, but temporary ripple configurations.

It also lets us:

- Model matter formation without invoking a Higgs field,
- Predict atomic decay from ripple misalignment, not random probabilistic tunneling,
- Understand black hole evaporation and element formation in stars as field-level reorganizations — not particle dance.

Summary Thought

Atoms aren't bricks of the universe. They are self-sustained field patterns — ripple nests held in feedback resonance. Their "mass" is memory, their "weight" is stability cost, and their "structure" is a story of survival in a vibrating, intelligent field.

When you hold a piece of matter, You are holding the most durable memory of ripple alignment nature can afford.

Chapter 6: Energy Is Not a Thing

Ripple Reconfiguration, Not Substance Transfer

What we call "energy" is not a stored fuel, but a change in alignment. There is no object being handed over — only field behavior shifting.

You've been taught to think of energy as something you can possess. Stored in batteries. Burned from fuel. Transferred through wires. Measured in joules, calories, kilowatt-hours.

But what if this entire view — elegant, practical, measurable — is fundamentally flawed?

What if there is no such "thing" as energy?

Not as an object. Not as a quantity passed around like cosmic currency. Not as an invisible substance waiting to be unlocked.

Instead, what we perceive as "energy" is simply this:

A change in the ripple configuration of the field —a shift in alignment, a distortion in memory, a consequence of interaction.

Nothing travels. Nothing is stored. Nothing is transferred.

A battery doesn't hold energy — it holds a readiness to misalign. A fire doesn't release energy — it undoes coherence. A wave doesn't carry energy — it ripples alignment across space.

This chapter will dismantle the myth of energy-as-substance and rebuild our understanding from a field-first perspective. We will see "power" not as stored potential, but as real-time reorganization of ripple structures. We will challenge thermodynamics, combustion, and even our deepest instincts about motion and heat.

Because if energy isn't a thing, then every law built on that assumption needs to be rewritten.

Let's begin.

The Substance Illusion

"Energy" as Object — A Historical Misunderstanding

The modern world treats energy like it's a commodity. We speak of *generating*, *storing*, *transmitting*, and *consuming* energy as if it's a tangible substance—a thing that flows like water from one place to another, measurable and manageable.

But this way of thinking, though useful in engineering and commerce, is conceptually misleading in the realm of fundamental physics.

It is not a mistake of mathematics. It is a metaphor gone rogue.

The Tank Metaphor: Energy as Stuff

From the Industrial Age onward, science needed a way to quantify motion and effort. To make machines predictable, power traceable, and work accountable.

Thus came the metaphor:

Energy is like a fluid in a tank — you have a certain amount, you use it, and then it's gone.

This mental model worked:

- Steam engines burned "fuel" (chemical potential) to produce motion (kinetic).
- Water behind a dam had "stored" gravitational energy.
- Batteries "held" electrical potential, waiting to be "released."

Each of these became analogies, and these analogies became laws.

Over time, energy became treated like a physical entity — an invisible liquid behind every observable process. Joules, kilocalories, kilowatt-hours — all units meant to suggest a countable essence.

But no one ever held an "energy molecule." No experiment ever revealed energy in isolation. It was always inferred, never directly found.

Kinetic and Potential Energy: Labels, Not Realities

Classical mechanics, through Newton and later Lagrange, gave us kinetic and potential energy.

- A moving object? It has kinetic energy.
- A raised object? It has gravitational potential.

But these were bookkeeping systems — ways to balance equations, not to describe *what energy is.*

UED exposes the core error:

These aren't two forms of stored "stuff." They are two different ripple configurations in the field.

Kinetic energy is not something *inside* a moving ball. It is the field's pattern of ripple displacement around the moving object.

Potential energy is not stored in the height of a rock. It is the readiness of the ripple field to shift if alignment changes (e.g., if the rock falls).

There's no storage. Only possibility — encoded in ripple positioning.

Calories, Joules, and the Packaging of Abstraction

The calorie started as a nutritional metaphor. A unit of heat — then converted into mechanical equivalents.

Joule, Watt, Electron-Volt — units to make energy feel real, accountable, economic.

And while essential for engineers, this packaging led to a conceptual trap: We began thinking of energy as a substance — a universal currency that can be accumulated, hoarded, or depleted.

This idea bleeds into everything:

- A power grid "delivers" energy.
- A food label shows "energy content."
- A star "produces" energy through fusion.

But none of these imply actual transfer of substance.
What's happening in each case is ripple realignment:

- Charges shift,
- Atoms destabilize,
- Fields reorganize,
- Perception reconfigures.

The Metaphor That Misled Physics

Physics embraced the metaphor because it made the math work. But in doing so, it:

- Hid the field dynamics behind numerical abstraction,
- Missed the opportunity to explore consequence-based causality, and
- Created conceptual myths that still linger in classrooms and research labs.

Here's the critical error:

We began mistaking dimensional accuracy for physical reality. Just because we can measure "10 joules" doesn't mean there is a thing called "10 joules" inside the system.

Energy is not a fluid. It is not a particle. It is not a force.

It is a description of how field configurations change —A label for consequences, not an ingredient.

From Storage to Shift — The UED Perspective

Under Universal Energy Dynamics, energy is:

- Not a conserved quantity moving through space,
- But a field pattern reorganizing itself to maintain local equilibrium.

There is no transfer — only transition. There is no loss — only ripple redistribution. There is no storage — only readiness to reconfigure.

The implications are massive:

- "Heat" is not stored energy — it's ripple chaos.
- "Work" is not energy exerted — it's ripple resistance overcome.
- "Power" is not fuel burned — it's alignment changed per unit ATT.

We must now leave the comfort of energy-as-stuff and move toward energy-as-difference — a living, dynamic, pattern-centric understanding of how reality evolves.

ΔRipple — The Real Face of Energy

Energy is Change in Field Alignment

At the heart of Universal Energy Dynamics (UED) lies a transformative realization:

Energy is not an entity. It is not a substance. It is a difference. Specifically, a difference in ripple alignment — a ΔRipple.

Every event you call "energetic" — be it a flash of lightning, the heat of a flame, the burst of a rocket, or the glow of a screen — is nothing more than a reconfiguration of the ripple field. Not a substance changing hands. Not an invisible juice being injected. But a pattern being disturbed, broken, reshaped, or restored.

Every Event = A Change in Ripple Configuration

Imagine the universe as a still pond of Structured Vacuum Energy (SVE). Now drop a stone — a disturbance.

That disturbance is ΔRipple — the real face of energy.

When two objects interact — one burns, one collides, one charges — you're not watching energy *flow*, you're witnessing the field adjust its internal configuration.

Energy, then, becomes not a *thing that moves*, but a state that shifts.

- A moving ball? Its ripples are offset in space-time alignment.
- A hot iron? Its internal ripple cycles are vibrating in chaotic micro-misalignment.
- A battery? A ripple structure holding *instability* in tension — ready to reconfigure.

In every case:

Energy = How much the ripple has moved away from stillness.

"More Energy" = More Ripple Complexity

When we say something has "more energy," what do we mean in UED?

Not that it contains more of something, but that its ripple structure is:

- More complex,
- More misaligned, or
- More ready to reconfigure.

This gives us a new grammar:

- A stretched bowstring = ripple configuration under tension.
- A running athlete = continuous ripple displacement through body-field coupling.
- A gamma ray = a sharp, high-frequency ripple break through the perceptual field.

High energy is sharp contrast in ripple states. Low energy is subtle variation. Zero energy is equilibrium — not absence, but perfect alignment.

Heat, Motion, Light, Charge — All Ripple Misalignments

The four horsemen of classical energy — heat, motion, light, and electric charge — are reinterpreted in UED as:

- **Heat**: Ripple misalignment at micro-scale. The "randomness" of particle motion is simply unstable ripple cycles attempting to restore balance.
- **Motion**: A coherent ripple displacement. The object doesn't "carry energy" — it holds a traveling ripple form.
- **Light**: Final feedback ripple before form dissolution. It's not traveling energy — it's the last echo.
- **Charge**: A ripple asymmetry — a standing imbalance in the ripple field that causes relational distortions.

None of these represent stored energy. Each one represents a ripple deviation from equilibrium — and hence, ΔRipple.

This shifts the question:

Not "Where is the energy stored?" But "What alignment is changing — and how sharply?"

Power Is Not Delivered — It's Felt as Realignment Pressure

Let's be radical:

Power isn't delivered. It's experienced.

Your body doesn't "receive energy" when eating food. It undergoes ripple restructuring from molecular breakdown — felt as temperature, motion, thought.

A motor doesn't "consume electricity." It confronts field resistance — and realigns magnetic domains to rotate.

Power plants don't "supply power." They orchestrate macro-scale ripple instabilities across transmission mediums.

This is why:

- You feel powerful when you're aligned and directed.
- You feel exhausted when ripple harmony is lost.
- Machines "burn out" not from lack of energy — but from excessive misalignment pressure.

Power, then, is not what something *has* —It's what something must *endure* to stay coherent during transformation.

ΔRipple: The True Currency of Physics

We no longer need to imagine tiny invisible boxes of "energy" being passed around the cosmos. We can now see the universe as a vast, intelligent field — whose every expression is a ripple consequence.

- Motion? Realignment over distance.
- Sound? Pressure ripple in medium.
- Explosion? Sudden collapse of ripple configuration.
- Life? Sustained ripple memory resisting entropy.

In UED, we don't ask how much energy something has. We ask:

What ripple difference is it maintaining — and what would it take to change it?

Because in truth:

All energy is ΔRipple. And all change is a story of alignment.

Stored Power? Or Delayed Realignment?

Reinterpreting Batteries, Springs, and Fuel

What does it mean to "store energy"? That phrase is repeated in physics classrooms, engineering labs, and even fitness apps. But in UED, we replace that phrase with something far more precise:

"Storing energy" is a misnomer. What's actually stored is ripple tension — a configuration of imbalance waiting for the right condition to realign.

In short, nothing holds energy like water in a tank. What exists is readiness to reconfigure. That readiness is encoded in the field — not locked inside objects.

Let's explore the big three: batteries, springs, and fuel.

Batteries: Field Configurations Waiting for Realignment

In classical understanding, a battery "stores chemical energy." But think beyond chemicals — what's really happening?

A battery is a ripple container — an artificially stabilized set of misaligned chemical states. Electrons are not little projectiles with "stored" energy. They are ripple interactions — suspended under field pressure created by chemical imbalance.

When a circuit is closed, it doesn't "release energy." It triggers a realignment pathway — a gradual collapse of the ripple misalignment stored across the battery's terminals.

Think of it like this:

- The voltage is ripple potential difference.
- The current is ripple readjustment rate.
- The heat and light at the endpoint are just perceptual feedbacks — visible consequences of alignment correction.

So, the battery is not a "box of energy." It's a contract of ripple instability — fulfilled when the circuit gives it permission to realign.

Springs: Geometric Ripple Tension, Not Compression Storage

A spring is a marvel of simplicity. Compress it, stretch it — and it "stores energy." Right?

Wrong.

In UED terms, a spring is:

- A geometrically encoded ripple configuration,
- Whose field alignment has been disturbed but remains self-recoverable.

When compressed or stretched, the atoms in the spring hold new local ripple tensions. The metal lattice maintains field rigidity, but it is misaligned from its rest state.

Let go, and the spring "releases" energy?

No — it realigns. The structure finds its minimum ripple stress state — its natural equilibrium. That "snap back" isn't a discharge of stored content — it's a field returning to coherence.

So, we should teach:

A spring doesn't store energy. It remembers misalignment — and seeks to forget it.

Fuel: Not Energy-Rich, But Misalignment-Rich

Fuel is often considered the ultimate storehouse of energy. But no hydrocarbon *contains* energy like a battery contains charge.

Instead, fuel is a ripple structure waiting to collapse.

Molecules like gasoline or coal are complex ripple nests — high in structural asymmetry and field imbalance. What makes them "combustible" is their tendency to dissolve into simpler, lower-energy ripple configurations.

Burning, then, is not the "release" of stored energy — it is the violent collapse of unstable ripple patterns. The resulting heat, light, and pressure? They are perceptual consequences of that collapse.

Think of fuel like a sandcastle built at high tide:

- Beautiful, delicate, and unstable.
- It takes effort (construction = ripple complexity).
- The incoming wave (triggering agent) causes a fast return to ripple simplicity — flattening the form.
- What you see — splash, wave, mist — is not substance escaping. It is feedback.

So, fuel isn't energy-rich. It's ripple-rich — packed with tension waiting to realign.

From Storage to Readiness — A New Vocabulary

Classical Term	UED Interpretation
Stored energy	Ripple misalignment tension
Potential energy	Readiness for alignment shift
Energy density	Ripple instability density
Energy release	Collapse of ripple coherence
Fuel efficiency	Rate of usable realignment feedback

By removing the illusion of storage, we free ourselves to ask better questions:

- What is the ripple complexity of this system?
- How stable is its configuration?
- What will it take to trigger realignment?
- What is the feedback consequence of that shift?

In this model, the universe is never "charged" or "empty." It is always in flux — somewhere on the spectrum of alignment, misalignment, and return.

Nothing Is Stored. Everything Is Waiting.

A dead battery is not "empty." It's aligned — too stable to do work. A stretched spring is not "loaded." It's tense — not for what it holds, but for what it remembers. A log doesn't "contain energy." It's carved in ripple layers that only fire can dissolve.

In UED, energy isn't what's stored. It's what's reconfigured.

What appears to "burn" or "release" is simply the visible echo of ripple patterns returning to rest.

Energy Transfer Is Field Update

No Flow — Just Ripple Rewriting

Let's be honest: the phrase "energy transfer" has fooled generations.

It paints a picture that feels intuitive — like something is *moving* from one place to another. A bucket passed from hand to hand. A stream flowing through a pipe. A charge sent down a wire.

But in the UED framework, this imagery collapses. There is no *flow*. There is no *hand-off*. There is no *thing* moving through a void.

There is only one phenomenon occurring:

The field updates itself — ripple by ripple, alignment by alignment.

Let's unpack this paradigm shift.

Energy Transfer in Electricity? Field Rewriting.

In classical physics, electrons carry energy along a conductor. Electricity flows. Current moves. Power transmits.

But what actually moves? Electrons themselves barely drift — only millimeters per second. Yet light bulbs switch on instantly.

This is not flow. It's ripple update. The electromagnetic field realigns across the circuit, triggering perceptual effects (heat, light) along the path.

There is no "delivery" of energy. There is only field response.

UED explains it this way:

- A change in one part of the field forces a neighboring ripple to adjust.
- This triggers a chain of local updates — not a transfer.
- Think dominoes toppling — not marbles flying.

Energy doesn't move. The field reconfigures.

Sound? Not Vibration Transfer — Field Disturbance Propagation

What about sound?

Textbooks say: "Sound is energy carried by vibrating molecules." That implies energy *moves* from source to ear.

But again, molecules don't travel across the room. They return to equilibrium. What travels is disturbance — not matter, not energy parcels.

In UED:

- Your vocal cords disturb the ripple alignment in the air's field.

- This disturbance propagates — each molecule's local ripple adjusts, then passes influence to the next.
- By the time it reaches someone's ear, the ripple pattern has been replicated — not moved.

So what was transferred?

Nothing. The airfield updated itself — under localized pressure. Your "sound" is a field memory, traveling across air through momentary ripple tension.

Radiation? Not Emission — Echo Collapse

Even in radiation — the most "emissive" of classical ideas — UED tells a different story.

A radioactive nucleus doesn't throw particles through space like a gun. It collapses a ripple structure, creating feedback that the field registers.

That gamma ray? It's not a photon flying through vacuum. It's the final ripple re-alignment from a decaying configuration — a last echo from a form that can no longer hold itself stable.

When we "detect" that radiation:

- We don't catch a particle.
- We read the field disturbance that just arrived — the memory of collapse.

Transfer? No. Field update at detection point? Yes.

"Work" = Overwriting Ripple Stability

In thermodynamics, work is defined as force over distance. In UED, we refine that:

Work is the cost of disrupting local ripple stability.

When a weight is lifted:

- The object's ripple configuration must resist gravity's alignment gradient.

- To lift it, your muscles — themselves ripple distorters — impose external alignment.

You don't push a thing up. You rewrite the ripple configuration that defines "down."

When a motor turns, when a current flows, when a planet spins — it's not transfer, it's realignment. Local patches of the field adapting to new conditions under tension.

Work, then, is not about movement. It's about misalignment effort required to overcome the field's natural coherence.

A Summary Table — Classical vs. UED

Classical View	UED View
Energy is transferred	Field is updated
Energy is stored	Ripple configuration is misaligned
Energy is emitted	Ripple coherence collapses
Energy is absorbed	Field re-stabilizes via new ripple state
Work is force x distance	Work is ripple disruption cost

The Ripple Net Always Updates Itself

Let's be clear:

The universe does not push packets. It realigns field conditions.

You don't transfer heat — you destabilize local ripple coherence. You don't transmit electricity — you induce field reaction across distance. You don't burn fuel — you dissolve high-stress ripple nests into simpler patterns.

Everything that "moves" is actually updating.

The ripple field is not a highway — it's a fabric of reactions.

There is no flow.

Only memory, effort, and change.

Implications for Thermodynamics

Heat, Entropy, and the Illusion of Loss

Thermodynamics was built on a story: that energy is substance, that heat is motion, and that entropy is inevitable decay.

This story has served well — until now.

UED reframes the narrative.

There is no "energy" being burned, lost, or used up. There are only ripple configurations changing — from coherent to chaotic, from tight to loose, from memory to noise.

Let's dismantle the old script and reassemble it through ripple truth.

Heat Is Not Motion — It's Ripple Agitation

In classical thermodynamics, heat is molecular velocity. The hotter something is, the faster its particles move.

But UED doesn't see particles. It sees ripple clusters — temporary alignments in a dynamic field.

So what is heat?

Heat is agitation in ripple alignment — the jittery instability of a field unable to stabilize.

Think of it like this:

- A cold object has stable ripple cycles.
- A hot object has frantic ripple activity, where coherence is repeatedly disturbed and re-formed.
- That "vibration" is not motion through space — it's oscillation in field memory.

Thus, temperature = ripple agitation level, not velocity.

It's not how fast matter moves. It's how unstable the alignment has become.

Entropy Is Not Disorder — It's Ripple Desynchronization

Entropy is often described as disorder — a vague term. Or as the "number of microstates," which sounds mathematical but feels hollow.

In UED, entropy becomes far more intuitive:

Entropy is desynchronization of ripple memory across the field.

Let's break that down:

- Coherent systems (like a crystal or living cell) have synchronized ripple cycles.
- When energy (realignment pressure) is introduced, it disrupts this coherence.
- The system doesn't break — it loses alignment.
- Over time, it becomes a soup of unsynchronized ripple activity — no longer able to respond as a whole.

That's entropy:

Not disorder. But loss of unified ripple behavior.

And entropy never destroys energy. It only scatters alignment beyond recognition.

Nothing Is Lost — Just Misaligned

Classical thermodynamics fears loss. "Energy dissipates." "Useful work becomes unavailable." "Entropy wins."

But UED says otherwise.

Nothing is lost. Nothing disappears. The ripple field simply spreads out its configurations.

Let's take examples:

- A hot cup cools down? Not energy loss — just ripple agitation absorbed by surroundings.
- Fuel burns in a car? Not destruction — just realignment of high-stress configurations into simpler ones, plus some ripple feedback (light, heat).
- A dead battery? Not depletion — just ripple configurations that no longer hold high reconfiguration potential.

UED reframes the "conservation of energy" law as:

The ripple field always balances its memory and misalignment — nothing escapes, only spreads.

Energy isn't conserved. Ripple potential is.

Work, Heat, and the Ripple Economy

What we traditionally measure:

- Work = external alignment imposed on the field.
- Heat = agitation from misalignment inside the field.
- Entropy = breakdown of synchronized response in the field.

All of this is ripple economy:

- Where "useful" means coherent enough to respond.
- Where "loss" means so spread out it can't reform easily.

It's not magic. It's not fate. It's not some mysterious arrow of time.

It's ripple consequence.

Final Realization

Thermodynamics doesn't need to be rewritten. It needs to be re-understood.

UED doesn't oppose the first or second law. It grounds them in field behavior:

- Energy is not burned.
- Entropy is not decay.
- Heat is not speed.
- Work is not force times distance.

All of these are different faces of one thing:

Field alignment under tension.

The universe doesn't lose. It simply reconfigures.

UED's Consequence-Based Law of Energy

There Is No Stored Energy — Only Ripple Availability

In classical physics, energy is treated like a quantity in reserve — as if the universe is a warehouse full of fuel tanks waiting to be tapped.

But in Universal Energy Dynamics (UED), this view collapses. Not because it's mathematically wrong — but because it lacks physical meaning.

UED reframes energy not as a thing but as a state of readiness:

There is no stored energy — only ripple availability, waiting for alignment triggers.

Let's unpack what this truly means.

Energy Is Not "There" — It Is "Ready"

When we say something "has energy," what do we mean?

- A battery "has" 5 volts?
- A ball at height "has" potential energy?
- A compressed spring "contains" stored energy?

These phrases are conceptually misleading. They suggest that energy is a thing present in space, waiting to be released — like water behind a dam.

UED sees it differently:

Nothing "has" energy. Everything sits in a specific ripple configuration that is ready to reconfigure under the right trigger.

A battery is not charged with invisible substance — it holds field tension. A spring is not full — it's misaligned in geometry. A ball at height is not elevated in energy — it's out of harmonic position in the ripple field.

Thus, "energy" isn't present — it's potential re-alignment.

All Action Is Field Shift — Not Energy Transfer

We say:

- A car engine "delivers power."
- A heater "releases heat."
- A moving object "carries kinetic energy."

In UED, all these are not flows but shifts:

Every action is a ripple update in the field — not something transferred, but something reorganized.

Imagine a still pond.

- Throwing a stone doesn't add water — it realigns it into waves.
- Light doesn't travel — it propagates as ripple feedback.
- Electricity doesn't flow like fluid — it triggers ripple oscillation across the field nodes.

So when we observe "power" or "work" being done, we're really witnessing a consequence of ripple misalignment being resolved — locally, not linearly.

Energy Is Constant — Experience Changes

UED holds that Universal Energy is a constant — not because it obeys a textbook law, but because structured vacuum energy (SVE) underlies all phenomena.

What changes is not how much energy exists, but how ripples are configured, how they interact, and how their misalignments manifest in observable form.

A fire doesn't destroy energy —it breaks ripple bonds and releases feedback (heat, light) as a field realignment cascade.

A human being doesn't consume food as energy substance — they trigger ripple resonance in biological configurations to support function.

Every change, from falling apples to exploding stars, is a change in ripple geometry, not an addition or subtraction from a universal fuel tank.

From Numbers to Consequences

Conventional physics tells us:

"Energy is conserved." "Energy can neither be created nor destroyed."

UED agrees — but with a twist.

It adds:

What matters is not the number — but the consequence.

Two systems may have the same "energy content" by joule count, but yield radically different ripple consequences — one may explode, the other may resonate calmly.

Therefore:

- Energy understanding must not stop at quantification.
- It must reach into field configuration and outcome.
- Not just how much, but how it behaves.

That's UED's Consequence-Based Law of Energy:

You do not measure energy — you align with its potential. You do not store energy — you configure ripples. You do not transfer energy — you trigger responses in the field.

This law doesn't replace conservation — it reveals its deeper nature.

Closing Insight

The entire energetic universe is not passing objects around. It is rebalancing ripple conditions — over and over again, in every event, every action, every experience.

What we call "energy" is just the ripples talking to each other.

There is no container. There is no handover. There is only the readiness of the field to change.

That's the new law. And it explains everything without inventing anything.

Chapter 7: Light Is Not Emitted

Ripple Feedback from the Last Alignment — Ending the Illusion of Radiating Objects

What if everything you were taught about light — was backwards?

We've imagined stars flinging photons, lasers firing particles, and light bulbs launching waves like bullets from a gun. We've grown up in a world where "light is emitted," as if something leaves an object, travels through space, and hits your eyes.

But under Universal Energy Dynamics (UED), that entire picture collapses into illusion.

Light isn't emitted. It's not a thing that moves. It's not a substance that travels. It's not a stream, beam, or bullet.

Light is ripple feedback — the *final perceptible echo* of a configuration collapsing, aligning, or shifting at the edge of your perception.

You never catch light in transit. You only see the last moment of coherence — the visible signature of a field realignment that already happened.

This chapter dismantles the entire object-radiation model. It redefines "illumination" not as the motion of particles, but as the resonance of observation. From Einstein's photon to Planck's quanta, from candle flames to gamma bursts — we'll reveal what's really being seen, and more importantly, what never moved.

You don't see because something reached you. You see because something aligned with you.

Let's end the myth of emitted light — and enter the ripple truth.

The Emission Assumption

From Flame to Photon — How We Built the "Light as Output" Myth

Since ancient times, humans have interpreted light through the lens of what they could *see* and *touch*. A burning flame gave off light. The sun lit up the earth. A glowing ember radiated brightness. Naturally, the assumption formed: light must be something coming out of these objects.

And that assumption never left us.

From early ray theories of vision in antiquity, to Newton's corpuscles, and on to Maxwell's electromagnetic waves, the notion of light as a thing in motion has persisted. Classical physics treated light like any other phenomenon: a product of a source, moving through space, interacting with matter.

Then came the quantum twist — the photon.

The photon was a patch, not a principle. Introduced to explain anomalies like the photoelectric effect, it painted light as a particle of energy emitted by excited atoms. Wave theory failed to explain why only certain frequencies could eject electrons. So Planck and Einstein offered a compromise — a quantum bullet carrying energy in discrete packets.

Suddenly, light had substance again. Something measurable. Something portable. Something that could be "emitted."

But this patchwork — the famous wave-particle duality — only deepened the confusion. It didn't resolve the contradiction. It rebranded it.

Quantum theory insisted light was both a wave *and* a particle, depending on how you looked at it. But what it never addressed is *why* light appears at all — and what it *really* represents in the fabric of existence.

Classical and Quantum: Both Assume a Source

Both frameworks — classical and quantum — make the same foundational mistake: They assume that light is emitted from an object.

- Classical: Light is an electromagnetic wave launched from vibrating charges.
- Quantum: Light is a photon particle emitted during energy transitions.

In both, an object emits, and light travels. This worldview requires light to move, occupy space, and take time — thus demanding a velocity, a medium, and a notion of direction.

But here's the UED correction:

Nothing emits light. Nothing travels. What we perceive as "light" is not transmission — it is a field-level feedback caused by ripple realignment.

UED Premise: Light Is Revealed, Not Emitted

In Universal Energy Dynamics (UED), all physical interactions are interpreted as ripple configurations within a structured field. Mass, motion, energy — all are forms of ripple alignment or resistance.

Light, in this model, is not a substance moving through space. It is the final observable effect of a ripple configuration reaching a threshold of alignment — one that crosses the ATT (Alignment Threshold Time) and becomes perceivable to a biological observer.

It's not that light "left" an atom and "arrived" at your eye. Rather, the field updated, and your mind synced with the feedback. What you see is not motion. What you see is resonance.

Why It Matters

This changes everything.

- Light is not proof of emission, but evidence of alignment collapse.
- There is no "source" — only synchronized field behavior.
- The emitter is an illusion; the revealed echo is the reality.

In short, we've been mistaking visibility for motion. We called it "light emitted," when in truth, it was light allowed.

UED doesn't merely critique emission theory — it overwrites it with ripple logic that restores coherence to the act of perception.

Ripple Feedback, Not Radiance

Perceiving the Final Echo — Not Emission, but Field Synchronization

If an object "radiates" light in the middle of an empty universe, and no observer exists to receive it — does light still happen?

In traditional physics, the answer is yes. Light is assumed to exist independently, racing away from its source, indifferent to whether it's ever seen.

But UED reframes that story — radically.

There is no light without perception, and no emission without ripple interaction.

Every "Light Event" Is a Consequence — Not an Activity

In Universal Energy Dynamics, the universe is not built from objects acting on each other. It is built from ripple fields interacting within alignment thresholds. Everything that appears to "happen" — from a falling apple to a glowing star — is a realignment of field structure.

This includes light.

- A "light event" is not an object sending particles into space.
- It is a ripple configuration collapsing or stabilizing, creating a moment of visible coherence.
- You see the result only if your perceptual threshold (your ATT) aligns with the ripple's feedback cycle.

So light is not emitted. It is revealed as consequence.

And more crucially: it is revealed only at the moment of alignment — never before, never after.

Ripple Collapse = Observable Coherence

When a ripple configuration in the field reaches a state where it can no longer maintain stable alignment, it undergoes collapse — not in the quantum mechanical sense of probabilistic reduction, but in the ripple dynamic sense of losing localized coherence.

This collapse emits nothing. Instead, it resolves the structure, leaving behind a final echo — a coherent flash or moment of perceptual clarity. That echo is what we call light.

It's not radiated. It's the last ripple you're able to perceive before the form vanishes or transitions into a less coherent state.

You don't see a wave traveling. You witness the result of an internal field negotiation — a synchronization between the collapsing ripple and your biological alignment capacity.

You Never See Light Travel — You See Alignment Points

You've never seen light in motion. You've only seen points — beginnings and ends. Glows. Flashes. Beams *starting* here and *ending* there.

But never the motion itself. Why?

Because light is not a journey. It's a relational event between field configurations — you and the source.

- When your field structure (your eye, your sensor, your mind) aligns with the feedback of a dissolving ripple — *that* is when light appears.
- There's no "stream" between the star and your retina.
- There's only a coherent update occurring at both locations in the ripple field.

This is why stars millions of light-years away can blink into your sky. Not because photons survived a multi-million-year trek, but because the field allows alignment across spatial tension — and reveals it *when* and *where* ATT permits.

Why This Dismantles the Emission Paradigm

This section erases the need for:

- Photon packets
- Emission spectra traveling through voids
- Velocity-based explanations for perception

Instead, UED offers:

- A field-dynamic logic where light is a feedback loop,
- A perception-based threshold (ATT) as the only true limit,
- And a redefinition of "seeing" not as reception, but as synchronization.

In essence:

Light is not something you catch. It's something you become coherent with — just long enough to call it "seen."

The End of Object-Based Radiation

From Projectiles to Perception — Light as Echo, Not Emission

The common intuition — and the cornerstone of classical and quantum physics — is this:

"Objects emit photons. These photons travel. We receive them."

This is the premise behind bulbs glowing, stars shining, lasers firing, and screens displaying. It's how we teach students and design experiments. But this narrative crumbles when examined through the lens of ripple dynamics in Universal Energy Dynamics (UED).

UED doesn't need an object to *emit* light. It only needs a ripple configuration to *complete* itself in the field.

Let's dismantle the old paradigm and show what actually happens.

Photons Do Not Leave the Object — They Are Perceived Echoes

The idea of photons — discrete packets of light leaving a source and traveling independently — was invented to solve a problem: how does light carry energy if it's also a wave?

The solution, though elegant for its time, was purely instrumental. It helped explain experimental results. But it did not explain what is actually happening.

In UED:

- There is no "photon" flying through vacuum.
- There is no particle carrying energy from A to B.
- There is only ripple reconfiguration.

When an event (a burning atom, a vibrating lattice, a hot object) crosses a ripple coherence threshold, it produces a field-wide adjustment. If your perception apparatus (eye, sensor, etc.) is aligned with that adjustment — you experience the echo.

That echo — not a traveling object — is what we call "light."

Lasers, Stars, Bulbs — All Are Ripple Conditions

From the red dot of a laser pointer to the glow of the sun, we've been taught to imagine things throwing light outward. But what these sources actually do is maintain field tension — pressure zones within the ripple medium.

Each of them is a condition, not a launcher:

- A laser is a device tuned to create precise ripple coherence, funneled through tight field geometry.
- A bulb filament vibrates ripple patterns into the surrounding structured vacuum — and feedback aligns across visible ATT windows.
- A star is a massive ripple turbulence machine, with zones of temporary field coherence constantly forming and collapsing.

What we perceive is not light shooting out, but the last ripple feedback that reached coherent visibility within our own field's perception range.

Light Is Not Emitted — It's the Last Ripple Feedback

Think of light not as an arrow fired from a bow, but as the final shimmer on the surface of a breaking wave. You don't witness the water's entire journey — you see the last peak of tension before it crashes into stillness.

Similarly, a light event is the final, perceivable echo of a ripple interaction.

- Not "radiation" from a body.
- But field resolution happening in real-time across ripple geometries.

When an object's internal configuration fails to maintain coherence, it causes the field to update — and the aligned observer perceives that update as light.

Blackbody Radiation Reinterpreted: Surface Ripple Thresholds

Blackbody radiation is often cited as the ultimate proof of light emission — a hot object glowing across a spectrum based on temperature.

But in UED:

- A blackbody is not "emitting" radiation.
- It is constantly undergoing ripple instability at its surface boundary.
- That instability crosses coherence thresholds at different ripple scales — producing a field-wide feedback spectrum.

So what we call Planck's curve isn't about photons emitted — it's about how often and at what ripple resolutions the field fails to stabilize the object's edge configuration.

More heat = more agitation = more ripple breakdown = broader echo spectrum.

This replaces particle emission with threshold-triggered ripple collapse.

The End of the Emission Paradigm

This shift is more than semantic. It's foundational.

It means:

- You cannot locate a photon.
- You cannot follow a "ray" of light.
- You cannot store or release light from objects.

You can only measure when and where your field aligns with a coherent feedback moment in the larger ripple system.

This ends the idea of radiating objects. It replaces it with a deeper principle:

The universe does not throw light. It responds with visible feedback when ripple tension realigns.

That's not just a new interpretation. That's a new physics of visibility.

Observing the End, Not the Journey

Light is Not a Traveler — It's a Terminal Alignment Event

We have spent centuries assuming that light travels.

From ancient theories of emission to Newton's rays and modern quantum photons, the consensus has always been: Light goes from here to there.

But in Universal Energy Dynamics (UED), this assumption is not just incomplete—it's false.

UED declares:

You never observe a journey. You only observe a field alignment that happens to include you.

Let's break it down.

Light Is Not a Traveler — It's a Terminal Alignment Event

Conventional physics insists on a transport model:

- A photon is emitted.
- It travels at speed *c*.
- It reaches an eye or a sensor.
- Observation happens.

But UED says: no photon leaves, no trajectory exists, and nothing crosses space.

Instead:

- A ripple destabilizes in one region.
- This causes a field-wide adjustment.
- When you, the observer, are aligned with this feedback geometry, you see what we call "light."

So what we perceive is not an object in transit, but a terminal event in our perceptual field.

There Is No In-Between — Just Observer-Field Synchrony

Ask yourself:

- Have you ever seen light halfway between a source and your eye?
- Can you ever observe a photon in flight?

The answer is no. All perception is instantaneous at the point of field coherence.

There's no middle state — only:

1. Before alignment (invisible),
2. Moment of alignment (perception),
3. After alignment (fade or collapse).

This aligns with quantum observations (e.g., wavefunction collapse) but goes further: UED replaces probabilistic magic with field geometry logic.

You don't track a wave or particle. You sync with a ripple echo.

Speed of Light Redefined — A Limit of Readability, Not Transport

Einstein's famous *c* — ~299,792,458 m/s — is treated as a transport ceiling.

But UED reframes *c*:

Not as a velocity of travel, But as the limit of how fast the field can provide observable feedback.

In simple terms:

- Light doesn't "move" at speed *c*.
- Field realignment becomes perceivable no faster than *c* allows.
- This is a perception rate, not a motion rate.

It's the upper limit of ripple readability — like a system refresh rate or data retrieval speed.

Thus, *c* isn't how fast something moves — it's how quickly alignment can happen at the edge of observability.

ATT (Alignment Threshold Time) — The True Visibility Window

In UED, what truly governs visibility isn't *c*. It's ATT — Alignment Threshold Time.

ATT is the minimum time required for a ripple configuration to:

- Complete a coherence cycle,
- Achieve field alignment,

- And become visible to the observer.

When:

- **ATT is above threshold** → object appears stable and visible.
- **ATT approaches minimum** → object appears to flicker, fade, or glow (light).
- **ATT falls below limit** → object dissolves into field (invisible or unobservable).

So, visibility isn't a measure of emission or trajectory. It's a product of timing, alignment, and ripple coherence.

That's why:

- We see glowing edges during vaporization.
- We see "light" in collapse zones.
- We can't see high-energy particles in motion — they break ATT limits.

Light Is the Field Saying: "This Is the Last You'll See"

When you "see light," what you're really witnessing is:

The field's final successful attempt to hold something visible before coherence is lost.

There is no journey. There is only the event of observability, and then — the disappearance.

You never observe motion. You observe transition.

Summary Takeaway:

- Light isn't emitted.
- Light doesn't travel.
- Light isn't a substance.
- Light is your synchronization with the final ripple echo of a disappearing field configuration.

That's why it appears suddenly, vanishes quickly, and always feels mystical.

Light Without Source — Implications

Why the Field Doesn't Need a Flashlight to Illuminate Reality

When we say, "light comes from a source," we mean well — but we're wrong. From bulbs and stars to lasers and lightning, our entire physics education revolves around one myth:

"Something must emit light for us to see it."

UED breaks this illusion at the root.

Because light, in ripple dynamics, is not *from* an object — it is *of* the field.

Let's examine what happens when we stop treating light as a projectile... and start seeing it as feedback coherence.

Light Can Appear Without a Classical Emitter

Phenomena That Defy the "Source and Beam" Story

You don't need a hot object to create "light." You don't even need a physical emitter in the conventional sense. What you need is field coherence that reaches observable threshold.

Take Cherenkov radiation, for example.

- A particle travels through a medium faster than light can be perceived in that medium.
- It emits a blue glow — not because it's *emitting light*, but because its movement disrupts the ripple field in such a way that alignment feedback occurs.
- The glow is a byproduct of broken field symmetry, not a beam shot out.

Or consider fluorescence:

- Molecules absorb energy and then "emit" light after a delay.
- But what's really happening is that their internal ripple states

collapse into new configurations — and the resulting coherence creates visible ripples that you interpret as light.

Even "seeing nothing" is revealing. When you stare into a pitch-black space but still perceive faint outlines, glimmers, or illusions of form — you are not seeing emissions, but momentary field alignments, echoes within your ripple-perception threshold.

In short:

Light is not always caused. Sometimes it's just consequence.

Shadows and Reflections — Not Rays, But Interference Maps

You were taught that shadows are cast when light rays are blocked. And reflections are "bounced rays" from smooth surfaces.

That model dies under UED.

Because:

- There are no rays.
- Nothing is shot from a bulb into your eye.
- What you see as a shadow is ripple cancellation — zones where incoming and obstructed field alignments don't meet your ATT (Alignment Threshold Time) requirement.
- A reflection is not "redirected light" — it's a secondary ripple alignment caused by phase coupling at a boundary.

You don't see because light bounces. You see because ripple configurations allow visibility at specific locations.

This radically redefines every optical diagram ever drawn.

Why "Light Years" Are Metaphorical, Not Mechanical

Let's talk space.

"Light-year" is defined as the distance light travels in a year. But if light doesn't travel, what's really being described?

In UED:

- A "light-year" is a field coherence range — the maximum

distance at which an alignment event remains perceptually possible from a distant source.
- The "speed of light" is not the speed of motion — it's the rate at which ripple coherence can trigger perceivable alignment.

So the universe isn't a big theater where photons are flung across billions of kilometers. It's a structured ripple field where distant coherence echoes are revealed to observers based on ATT windows.

In this sense:

"Light-year" becomes a temporal-perceptual metaphor, not a hard mechanical path.

End of the Flashlight Metaphor — Welcome to Ripple Realism

Let's retire the torch. The idea that light "shines from here to there," illuminating objects like a stream of water — is dead.

There's no flashlight. There's no source or target. There's only the ripple field, constantly configuring and dissolving.

- Sometimes, that configuration becomes visible.
- Sometimes, it collapses before feedback aligns.
- What we call "light" is that moment of success — when field coherence reaches the observer.

It's not a beam. It's a blip.

A synchrony. A flash of alignment. No emission. No transport. No stuff.

Just real-time field negotiation.

Welcome to Ripple Realism — where seeing isn't believing...It's *being aligned.*

Chapter 8: The Echo of the Dissolved

What Fire, Explosions, and Decay Really Are

You never see an object vanish. What you witness is its final protest — a flicker, a flash, a heatwave — the last ripple escaping from what once held shape.

A flame is not fire consuming matter. It is form letting go, coherence collapsing into chaos, and the field struggling to stabilize.

We call it heat. We call it light. But all we ever perceive is the echo of a dissolved alignment.

The explosion is not a violent push outward. It is a rupture in ripple continuity, where structured energy can no longer hold the pattern — and the field reacts violently to restore balance.

Decay? It's not death. It's the slow forgetting of a structure that no longer resists entropy. The ripple memory weakens. What once stood firm now *seeps back into field neutrality*.

In this chapter, we abandon the myth of "energy release" and "matter breakdown." Instead, we dive deep into UED's truth:

"What burns, breaks, or disappears... doesn't leave. It just stops being alignable."

You don't see destruction. You feel the consequence of disintegration — and mistake it for an event.

Welcome to the true story of fire, explosions, and decay. This is not combustion. This is the ripple aftermath of the once-coherent.

The Moment of Collapse — Not Destruction, But Dissolution

We've long believed that when something burns, explodes, or decays, it's being destroyed — that material is breaking down into pieces or converting into energy. That myth has shaped entire scientific doctrines, from chemical reaction models to nuclear fission equations.

But Universal Energy Dynamics (UED) exposes a deeper reality: Nothing is destroyed. There are no "pieces." There is only coherence lost — and what we perceive is the echo of that moment of dissolution.

The Myth of Destruction: A Legacy of Particle Thinking

Traditional science teaches us that objects are made of particles, and when these particles break or rearrange, we get heat, fire, explosions, or decay. We are told that bonds are broken, atoms are released, and matter "converts" to energy.

But this assumes that something *solid* is breaking. It assumes matter is a container, and energy is a fluid that leaks out when the container cracks. This metaphor, elegant in chemistry textbooks, completely misrepresents the field-based reality.

UED's Reinterpretation: Collapse of Coherence

Under UED, everything that appears "material" is just a ripple configuration — a temporarily stable pattern of alignment in the field. The object you see is not "stuff" held together by bonds. It's a standing ripple held in harmonic resonance within Structured Vacuum Energy (SVE).

When fire happens, it's not matter releasing energy — it's alignment unraveling. The ripple configuration becomes unstable, collapses, and in doing so, generates turbulence in the field. That turbulence becomes visible as:

- **Light** (the final feedback of visible alignment)
- **Heat** (the agitation of surrounding field coherence)
- **Sound** (pressure fluctuations as field waves adjust)
- **Smoke** (residual misaligned fragments trying to stabilize)

Combustion and Decay: Different Speeds, Same Collapse

Fire and explosions are fast. Decay is slow. But the underlying phenomenon is identical:

Collapse of ripple memory into simpler, lower-tension field configurations.

- Fire is rapid de-coherence.
- Explosion is chaotic redistribution.
- Decay is gradual ripple leakage from a once-stable node.

You never witness matter being "broken." You're simply perceiving the symptoms of lost alignment. The flame is not "leaving" an object — it's the recoil of coherence collapsing.

You Perceive the Echo, Not the Fall

What we call "seeing something burn" is nothing more than our perception catching the final moments of a configuration dissolving. We mistake the echo for the event. We mistake the turbulence for the transition. And we invent explanations — combustion, oxidation, fusion — to label something that's not fundamentally understood.

But UED gives us the language to see it clearly:

All energetic events are not about force or flow. They're about ripple re-alignment. And collapse is just the moment when alignment can no longer hold.

Fire Is Not Burning — It's Ripple Rewriting

When we observe fire, we think of it as *burning* — a release of energy, a transformation of substance, a consumption of matter. But this is a metaphor inherited from a surface-level view of reality, built upon classical physics and chemical reaction models.

Under Universal Energy Dynamics (UED), fire is not burning — It is ripple rewriting. It is not a transfer of substance. It is the collapse of coherence and the realignment of the field.

Flames Are Not Emitted — They Are Turbulence Made Visible

A flame is not a "thing" that comes out of a burning object. It is a zone of ripple instability, where the structured coherence of matter fails and the resulting turbulence becomes visible to observers whose perceptual thresholds align with that instability.

In ripple terms:

- The flame is where coherent ripple configurations fall apart.
- The flickering motion is turbulence — not fuel particles dancing, but wavefronts scrambling for equilibrium.
- The shape and color of the flame represent the degree of misalignment and field reconfiguration.

There is no entity "leaving" the object. There is only a reversion of complex structure into chaotic field states.

Heat, Light, and Smoke: Echoes of Ripple Breakdown

Each "output" of fire is a different perceptual symptom of ripple disorder:

- Heat is field agitation — as ripple zones around the object get scrambled, nearby coherent zones are disturbed and reconfigured.
- Light is the last visible ripple feedback — not an emission, but the final echo of structure losing form.
- Smoke is what we call residual ripple fragments — misaligned packets no longer stable, attempting to re-enter field equilibrium.

These are not "releases" of stored substances. They are manifestations of a realignment process, as field memory resets to a less organized state.

Burning Is Not Emission — It's a Field Rebalancing

In classical science, fire is a chemical reaction — an exothermic process that "releases" energy by breaking bonds. But what if there are no real "bonds" — only field alignments?

Then what we call "burning" is actually:

A ripple configuration collapsing, and the field rearranging itself to accommodate the loss of structure.

UED says:

- Nothing is emitted.
- Everything is reconfigured.
- Energy is not freed — it is redistributed across the field in simpler, less coherent waveforms.

The fire isn't consuming. It's rewriting.

What Disappears in Fire Is Not Matter — It's Form

We believe a log turns to ash. That something is "lost" in the process — the rest escapes as heat and smoke. But the matter didn't leave. The substance didn't go anywhere.

What vanishes is the nested ripple alignment that once held the shape of the log.

Fire is a ritual of deconstruction. It's not violence. It's entropy, made visible.

There is no transfer. There is no flow. There is only the end of configuration, and the onset of turbulence.

Explosion = Sudden Ripple De-coherence

What Looks Like Force Is Just Field Panic

We've long believed that explosions are violent ejections of matter and energy, propelled by extreme internal pressure. But this is merely a visual narrative imposed on a much deeper, field-based phenomenon.

From the lens of Universal Energy Dynamics (UED):

An explosion is not a burst — it is a collapse. Not of substance, but of ripple coherence.

What we call an "explosion" is a sudden failure in field alignment — a ripple decoherence event that overwhelms the stability of the surrounding field and forces a chaotic readjustment.

The Shockwave Is Not Matter — It's Realignment Panic

We imagine a shockwave as a ring of gas or debris pushing outward with brute force.

But under UED:

- The shockwave is not material propulsion.
- It is a ripple spike — a rapid chain of alignment failures cascading through the field.
- Each zone of the field, encountering a sudden loss of coherence from its neighbor, reconfigures violently — and this collective chain becomes the "blast wave."

Think of it as a domino effect of field panic — not a wall of matter being shoved outward, but a wall of field restructuring rolling across space.

Sound, Heat, Flash — Stress Signals of a Field in Distress

What we perceive as:

- Sound: compression of ripple cycles causing acoustic realignment in the medium.
- Heat: ripple agitation — surrounding areas being forced into higher misalignment states.
- Flash: the final visible ripple feedback of the breaking form — just before coherence is lost.

None of these are "released" by a central substance. All are symptoms of field-wide ripple disturbances.

UED reveals:

These are not outputs. They are side-effects of field instability.

Explosion = Collapse Cascade

In classical physics, we use the word "detonation" to suggest combustion or sudden expansion. But that's a mechanical myth. What actually occurs?

A high-energy material (say TNT or uranium) is a highly strained ripple configuration. It's not storing "energy" — it's precariously balanced. Once the first misalignment happens, it can't self-correct — so the whole structure ripples apart.

- UED calls this a ripple collapse cascade.
- Each unit of the structure destabilizes the next.
- The environment is forced to compensate — not by receiving matter or energy, but by reorganizing ripple patterns violently.

Thus, the explosion is not pressure pushing mass —It's the field trying to restore balance after catastrophic misalignment.

Redistribution, Not Release

One of the biggest myths is that explosions "release" energy. But from a UED perspective:

There was no energy stored — there was only ripple tension.

There is no energy released — only field realignment.

In reality:

- High-energy explosions don't eject power into space.
- They force surrounding zones to adapt to a sudden ripple vacuum.
- What seems like forceful expulsion is actually the field redistributing imbalance at extreme rates.

And when that adjustment outruns perceptual thresholds? We see chaos. We hear thunder. We feel heat.

But nothing has moved — except meaning across a stressed field.

Decay Is the Field Letting Go

Why Things Don't Rot — They Dissolve into Simpler Field States

When we think of decay, we imagine rot, leakage, breakdown—a slow disintegration of matter, atom by atom. In biology, it's flesh decomposing. In chemistry, it's unstable molecules breaking down. In nuclear physics, it's radioactive material "emitting particles." All of this assumes something is being lost—as if pieces fall off a thing and fly away.

But under Universal Energy Dynamics (UED), that entire framing collapses.

Decay is not destruction. It is a ripple transition. Not something falling apart, but something being let go by the field.

Radioactivity, Chemical Decomposition, or Organic Decay

They're all the same ripple principle—just different scales.

Whether it's:

- A radioactive nucleus,
- A rotting apple, or
- A rusting nail...

The underlying event is not a loss of particles, but a loss of coherence.

What does that mean?

- Every atom, cell, or molecule is a ripple configuration, held together by nested alignment cycles.
- Over time, under stress or environmental triggers, these alignments slip.
- Not all at once, but in phases — creating instability zones.
- These zones no longer resonate with the surrounding field and begin to blend back into ambient states.

Hence decay is not a passive breakdown. It's an active reversion to simpler field states.

Atoms Don't Rot — They Desynchronize

Let's talk nuclear decay. The standard model says unstable nuclei emit particles (alpha, beta, gamma) and transmute into other elements.

UED says:

- No particle "leaves" the atom.
- What's actually happening: The ripple nest that makes the nucleus falls out of alignment with its own sustaining field pattern.
- When this happens, the field reconfigures that spot.
- What you observe as "radiation" is the echo of ripple realignment, not a particle flying out.

Even "decay chains" are just progressive ripple simplifications—each step stabilizing in a new pattern until it no longer triggers ambient field turbulence.

Biological and Chemical Decay — Same Pattern, Different Scale

- A dead leaf or rotting fruit follows the same logic.
- The complex biological ripples holding the tissue together begin to lose phase synchrony.
- Microbes and oxidation just accelerate this field desynchronization by introducing ripple interference.
- Eventually, the structure cannot maintain itself. It dissolves—not into pieces, but into field-level randomness.

From UED's lens:

You don't observe parts falling off — you witness a ripple nest losing identity.

Decay Is Perception of Ripple Turbulence

What makes us see, smell, or feel decay?

- It's not the object changing. It's our field interacting with its unstable ripple cycles.

- The odors? They're turbulent ripple echoes our biology reads as "unpleasant."
- The discoloration? A change in feedback pattern of the ripple structure.
- The texture change? A difference in resistance due to ripple incoherence.

So, when we say something is "rotting" — what's really happening is:

The field is letting go of a previously coherent configuration, and your own field is picking up the static.

Closing Insight

Decay is not death. It is ripple simplification. It is not matter vanishing, but identity dissolving.

No substance escapes. Only patterns dissolve — and the field resumes its calm.

Echoes Are the Only Reality We Perceive

Why We Never Witness Destruction — Only Its Aftershock

When something "disappears" — a flame burns out, a building collapses, a star explodes — we instinctively assume that we saw the event itself.

But did we?

No.

Under Universal Energy Dynamics (UED), we perceive only the ripple echoes — not the event, not the object, not the act of destruction. Just its *aftermath* in the field.

Let's explore this profound perceptual truth.

Dissolution Leaves Ripples, Not Remains

In classical thinking, destruction is a thing:

- Matter gets smashed

- Particles break off
- Debris is created

But in UED:

- What breaks is alignment, not matter.
- The object's ripple configuration falls apart — and in its place, field turbulence appears.
- That turbulence becomes heat, sound, light, or smoke — which is what we perceive.

So, when a structure dissolves:

What lingers is not its *material essence*, but its echo — its ripple resonance collapsing into ambient noise.

Examples of Echo Perception:

Let's look at common examples where we believe we're seeing "the thing", but we're actually just seeing its echo:

- **Heat shimmer**: You don't see heat — you see ripple agitation in the field distorting light coherence.
- **Smoke**: Not substance in the air, but a visual pattern of ripple incoherence left by combustion.
- **Sound**: A cascading pressure variation caused by ripple displacement — not something "traveling through air."
- **Smell**: Micro-vibrational turbulence stimulating receptors — the ripple, not the rot.
- **Memory**: A stored alignment trace in your brain's ripple field — not the event, but the record of its resonance.

Every one of these is a ripple echo, not a "thing."

There Is No Direct Experience of Destruction

This is a radical insight.

We never truly experience *destruction*. We only experience the consequences of collapse in our field.

Just like we never see light "travel," we never perceive a thing breaking — we perceive the ripple change that follows.

- The crash is an audio feedback
- The flash is a coherence spike
- The debris is a stable reformation, not proof of destruction

Our sensory world is a field of echoes, each one telling us what just stopped being aligned.

Object Permanence Is an Illusion

In traditional physics and psychology:

- Object permanence means things exist even when we don't see them.
- It's considered a hallmark of intelligence and cognitive development.

But UED flips that.

Objects never "exist" as things — they are persistent ripple configurations. Once those configurations dissolve, what's left isn't a thing disappearing — it's a ripple trace that slowly loses structure.

Hence, permanence is not physical — it is perceptual inertia. We assign continuity to ripples after the form is already gone.

Why Ripple Traceability Replaces Substance Tracking

In UED, there is no need to track objects through space.

Instead, we follow field updates:

- Where did the ripple begin to desynchronize?
- What sequence of echo events followed?
- How did coherence degrade into trace?

This allows us to:

- Reconstruct cause not by looking for particles, but by reading field footprints.
- Diagnose events not by matter analysis, but by ripple interference mapping.

This is the forensic science of UED — ripple archaeology.

Final Insight

You live in a world of consequences, not things. You see ripple echoes, not events. You remember patterns, not substances.

Everything you "know" is the echo of something that was once aligned. Reality, as perceived by consciousness, is the field whispering what just let go.

Implications — Fire as Final Feedback

Why Flames Don't Emit — They Reveal the End of Alignment

The poetic saying, "Everything burns to nothing," carries more scientific truth under UED than traditional physics dares to accept.

In classical thinking:

- Fire is an emission.
- The flame releases light and heat.
- Burnt = consumed.

But through the lens of Universal Energy Dynamics (UED), fire is not a process of releasing stored substance — it's the final feedback of ripple coherence dissolving.

Let's reframe the phenomenon, not as destruction, but as the last ripple visible before an object exits perceptual alignment.

The Last Light — Ripple Collapse, Not Radiation

When you stare at a flame, you're not witnessing something being emitted.

You're witnessing:

The final flickers of alignment failure — the very last ripple collapse visible to your field of perception.

The light of fire is not a product — it's a farewell.

- It is the last coordinated ripple that your sensory field can register before the form dissolves.
- There is no emission *from* the object.
- There is only feedback from the ripple's collapse, like the final splash before water settles.

Fire doesn't shine. Fire exits.

"Nothing Left" = No More Echo

Once the ripple coherence is gone:

- There's no field disturbance to register.
- No feedback signal to perceive.
- No alignment left to trace.

That's what we call:

Burnt out.

Not because energy ran out. Not because matter is "used up." But because the coherent ripple architecture has dissolved beyond sensory threshold.

There is nothing left to see not because the object vanished, but because the field no longer reflects it.

Burnt = Alignment Erasure

This is the core UED insight:

Fire doesn't consume matter. Fire marks the point of irreversible de-alignment.

To "burn" is to reach a threshold where:

- The ripple configuration can no longer sustain.
- It breaks into unstable forms.

- These unstable ripples spread out, fragment, interfere — and eventually flatten into ambient noise.

This is what the flame shows you — not an emission, but a dissolution signal.

Why the Flame Is a Boundary, Not a Substance

The flame itself:

- Has no "inside."
- Has no consistent shape.
- Is a moving alignment boundary, constantly forming and vanishing.

In UED, a flame is:

A threshold effect — a visible edge of ripple coherence rapidly failing.

The dancing motion? That's not "fire behavior." It's ripple instability, flickering because feedback isn't stable — because the structure is in its final moments.

That's why it's mesmerizing. You are witnessing the precise tipping point between existence and disappearance.

Why the Fire Ends When There's Nothing Left to Perceive

There's a moment in every fire when:

- The flame dims,
- The crackle stops,
- The heat dissipates,
- The glow dies...

And suddenly, it's dark.

What happened?

From the UED view:

The field no longer contains aligned structures capable of creating coherent feedback.

The final feedback has passed.

The object didn't finish burning. It completed dissolution.

There is nothing "to burn" anymore because there's nothing left to collapse in a visible way.

Conclusion: Fire as a Reminder of Perception's Edge

Fire teaches a profound UED lesson:

- You never see the event.
- You never touch the core.
- You never grasp the "thing."

You only witness the edge — the boundary where alignment dies and perception ends.

Fire is not fuel burning. It is form unraveling — and your senses catching only the brief, final echoes before the field returns to stillness.

This is why UED calls fire a consequence, not a cause.

It is not emission. It is not substance in action. It is the feedback signature of form leaving the field.

Key takeaways for Part II

1. Mass Is Not a Substance

- Mass is not "stuff." It's not a property of material—but a consequence of ripple coherence in the field.
- Weight and inertia arise from resistance to ripple displacement, not gravitational attraction or hidden mass.
- Density is alignment complexity, not particle tightness.
- What we perceive as "massive" is simply energetically stable ripple architecture resisting change.

2. Energy Is Not a Thing

- Energy is not stored fuel — it's the degree of field reconfiguration between ripple states.
- The myth of "stored power" in batteries, springs, or fuels collapses under UED — they hold potential alignments, not quantifiable "chunks" of substance.
- ΔRipple (change in alignment) is the only meaningful definition of energy.
- There is no transfer, only local field updates — energy doesn't move, the field realigns.

3. Light Is Not Emitted

- Light is not launched — it is a feedback effect of the last

coherent ripple alignment.
- There is no "photon leaving the object." The observed light is the final perceptible echo of a dissolving interaction.
- Whether stars, lasers, or bulbs, no object emits light — it reflects the ripple threshold collapse to the observer.
- Speed of light is not a velocity but a limit of perceptual coherence.

4. The Echo of the Dissolved

- Fire, explosions, and decay are not about matter breaking apart, but ripple coherence collapsing.
- Flames, flashes, and sounds are turbulent feedback, not substance leaving a surface.
- Explosions are ripple collapse cascades — not mass ejections.
- Decay is ripple desynchronization, not molecular death.
- All you ever perceive is the echo of something leaving alignment — never the thing itself.

Core Insight of Part II:

Reality is ripple-based, not object-based. What we call "mass," "energy," and "light" are just temporary alignments in a universal field, and what we perceive is never the origin — always the consequence.

Part III: Rewriting the Laws

Chapter 9: $E \neq mc^2$ — What It Should Have Been

Introducing the Realignment Equation

$$E_r = \mu_0 \rho_r \varkappa^2 (1 / ATT_0) \Delta R$$

(Chavan's Realignment Equation – The Field-Based Alternative to $E = mc^2$)

For over a century, a single equation stood like a monument over modern physics—$E = mc^2$. Simple. Elegant. Undefeated. So much so that it was rarely questioned, let alone challenged. We accepted it not as insight—but as gospel. Energy is mass. Mass is energy. Multiply by the square of light speed and you get the ultimate conversion rate between matter and meaning.

But what exactly is being converted?

What *is* mass? What *is* energy? And why should the speed of light—something never explained, only assumed—be the sacred glue binding them?

What Einstein gave us was a bridge—but no explanation for what walks across it. His equation passed the unit test, but failed the causality test. It *balanced dimensions* but *ignored the dynamics*. It told us what could happen, but never how or why it does.

That's where we begin again.

Not with a shortcut. Not with dimensional acrobatics. But with a consequence-based law grounded in ripple dynamics.

The Realignment Equation is not a rebranding of $E = mc^2$. It is a replacement.

Where Einstein saw substance turning into energy, we see something far more honest: Ripple configurations dissolving into perceptual consequence. No hidden fuel. No "stored" power. No imaginary particles flinging themselves across empty space.

Instead, we look at:

- How stable a ripple is (ρ_r)
- How complex its architecture is (κ^2)
- How fast its coherence can respond ($1 / ATT_0$)
- And how far the field must realign (ΔR)

This chapter does not revise Einstein's work. It liberates physics from it.

Because the universe isn't made of equations—it's made of alignment, and the cost of breaking it. Let's rewrite the most iconic equation in history — and finally make it tell the truth.

The Failure of $E = mc^2$

Why Dimensional Elegance Cannot Replace Physical Truth

It is time we confront a difficult but necessary truth: $E = mc^2$ has failed—not as a mathematical identity, but as a physical explanation.

At first glance, the equation is seductively simple. It suggests that energy (E) and mass (m) are interchangeable, with the square of the speed of light (c^2) acting as the conversion constant. It *looks* complete. It *feels* powerful. And it carries just enough symbolic weight to pass as divine insight.

But its dimensional beauty hides its conceptual poverty.

Let's unpack this:

- **E** is a measure of energy—but of what kind? Is it kinetic? Potential? Thermal? Electromagnetic?
- **m** is mass—but what *is* mass? Is it substance? Resistance? Field property? Is it static or dynamic?

- c^2 is the square of the speed of light—a velocity squared. But why squared? Why light? And why does this velocity suddenly become the universal currency of conversion?

There's no causality here. No process. No physical narrative.

The equation suggests equivalence but explains nothing about how energy emerges, why mass resists motion, or under what condition one becomes the other. It doesn't tell us *when* energy is released, *how* it propagates, or *why* certain forms of matter are more energy-rich than others.

And this is not just a philosophical gap—it's a scientific void.

In the world of Universal Energy Dynamics (UED), these oversights are unacceptable. Physics must answer why and how, not just how much. Without ripple mechanics—without the field-level architecture that actually defines mass and energy—$E = mc^2$ becomes an ornamental formula, not a foundational law.

It's like declaring, "Money = Gold × $Trust^2$" without ever explaining the banking system.

It's time we stop worshipping elegance and start demanding consequence. We need a framework that *doesn't just fit*, but unfolds the process. That shows us not just that energy exists, but how it arises from ripple reconfiguration, why it behaves differently across contexts, and what its release actually means in terms of field alignment.

$E = mc^2$ was a placeholder. A necessary historical bridge. But the bridge is now a blockade.

The future of physics begins when we stop solving for "E" and start listening to the ripples that define it.

Chavan's Realignment Equation — Origins and Meaning

Why Energy Must Be Measured in Ripple Terms, Not Mass Illusions

To replace a cultural monument like $E = mc^2$, one must do more than tweak variables. One must rewrite the very foundation of measurement—from substance metaphors to field realities.

That's what Chavan's Realignment Equation sets out to do.

$$Er = \mu_0 \cdot \rho_r \cdot \kappa^2 (1 / ATT_0) \Delta R$$

This is not just a formula. It's a physical map of ripple transformation. Each term holds not symbolic value but field-based consequence. This is energy as it *truly is* under the Universal Energy Dynamics (UED) framework: not a substance, but an unfolding of ripple configuration.

Let's decode it, term by term:

μ_0 – Baseline Field Constant

This is not borrowed from electromagnetism as a lazy parallel. In UED, μ_0 represents the universal field permeability—the ease with which ripple structures can form, interact, and transmit alignment through the vacuum field. It's the "grain" of the structured vacuum—a measure of how resistant or receptive the field is to change.

Think of μ_0 as the inertial softness of the vacuum. It sets the default difficulty level for ripple evolution.

ρ_r – Ripple Density

This is not mass density in the Newtonian or relativistic sense. ρ_r captures the concentration of ripple coherence within a field node—how tightly interlocked and energetically complex the ripple interactions are in a given configuration.

A high ρ_r doesn't mean "heavier." It means more ripple relationships to realign, hence greater potential consequence when that alignment breaks or shifts.

κ^2 – Structural Complexity Coefficient

If ρ_r tells us how much ripple is present, κ^2 tells us how intricately it's woven. This squared term reflects topological interdependence: loops within loops, nested harmonics, feedback coupling, fractal layering.

Why squared? Because complexity grows exponentially as the number of ripple interactions rise in tightly bound systems.

A sandpile has ripple mass. A brain has ripple *architecture*. That's the κ^2 factor.

1 / ATT_0 – Inverse of Alignment Threshold Time

ATT_0 defines the minimum time required for coherent ripple perception by a conscious observer or sensing system. Inversely, 1 / ATT_0 expresses how ready the configuration is to be *felt*, *read*, or *expressed* as energetic change.

Lower ATT = quicker ripple response = higher energetic availability. So energy isn't speed-dependent—it's alignment-ready-dependent.

This term replaces the frozen speed-of-light constant (c^2) in $E = mc^2$ with something far more elegant: a field-contextual limit.

ΔR – Realignment Distance

This is the core event. The change in ripple configuration—either through collapse, phase shift, interference, or reconfiguration.

ΔR is not spatial. It's topological—a measure of how *far apart* the initial and final ripple configurations are in alignment space.

A small ΔR? Minor energy ripple. A large ΔR? Explosion, phase shift, transformation.

Together: A Law of Field Consequence

Each term does not "measure" a thing. It captures readiness, structure, memory, and change—everything that *actually happens* in the field during an energetic event.

This equation:

- Doesn't talk about mass
- Doesn't require particles
- Doesn't assume emission
- And doesn't glorify constants that are contextless.

Instead, it quantifies ripple reconfiguration.

It gives us a consequence-based definition of energy: Not "how much substance is converted," but "how deeply the field had to realign to sustain or dissolve a coherent form."

In short:

Energy = Ripple Response x Complexity x Field Readiness

This is no longer physics of assumptions. This is physics of actual field behavior. The ripple is not a metaphor — it is the medium, the message, and the measurement.

And with this equation, energy finally regains its meaning.

What the Equation Means Physically

From Scalar Illusion to Structural Causality

Physics, at its worst, has been guilty of flattening behavior into numbers. Equations like $E = mc^2$ gave us elegant shortcuts, but they also gave us conceptual amnesia—the loss of any *why*, *how*, or *when*. Energy became a scalar abstraction, disconnected from the field that birthed it.

Chavan's Realignment Equation flips this.

$$E_r = \mu_0 \cdot \rho_r \cdot \varkappa^2 (1 / ATT_0) \Delta R$$

This is not a formula to "calculate" energy in isolation. It's a framework to understand energy as a physical consequence—a byproduct of ripple realignment under Universal Energy Dynamics (UED). Every term reflects what the field does when something appears, shifts, dissolves, or transforms.

This is not symbolic math — it's causal mapping.

Let's decode its physical significance.

Not a Scalar Shortcut — A Structural Diagnosis

The equation doesn't estimate how much energy is "inside" an object. It reveals what had to happen *in the field* for that object or event to be *possible*. It's like reading the tension report of a bridge, not just its weight.

In other words:

Energy is not a thing you have. It's a consequence of what your structure demands from the field.

Energy = Ripple Strain, Not Mass Conversion

There is no substance being converted here. What we call "energy" is the strain on the field to maintain, shift, or resolve a ripple configuration.

Let's break down each term and how it contributes to that strain:

1. Ripple Density (ρ_r)

Think of this as how much the field is already invested in holding a pattern together. A high ρ_r = more compact ripple cycles = more effort needed to disturb or preserve it.

In physical terms:

- A tightly packed crystal.
- A neutron star.
- A well-trained neural network.

All represent high ripple density. More density means more energetic consequence if realignment occurs.

2. Structural Coherence (κ^2)

This reflects the interdependence of ripple elements. Like a spider web or a musical chord — the more intricate the structure, the harder it is to adjust just one part without affecting the whole.

κ^2 is where complexity becomes cost.

You don't just move mass — you tug at a whole pattern. So the energy isn't stored — it's locked in relationships that resist or amplify change.

3. Alignment Threshold Time ($1 / ATT_0$)

This term defines how ready a structure is to show change. Low ATT = quick ripple reaction. High ATT = sluggish, stubborn coherence.

The inverse (1 / ATT_0) tells us how fast that configuration can *reveal* energetic transition. It is not a clock tick — it's the coherence refresh rate.

In practice:

- A fast-reacting molecule.
- A photon-generating event.
- A rapid combustion chain.

All happen when ATT_0 drops — and energy appears as snap ripple correction.

4. ΔR – Realignment Distance

This is the actual shift the structure goes through. It measures not space, but ripple state difference—how far apart two configurations are in alignment topology.

Even if the structure is dense and complex, no energy emerges unless there is realignment. ΔR is the trigger, the actual event, the *change* we later misname as "energy released."

So, Physically Speaking:

- Nothing is emitted.
- Nothing travels.
- Nothing is stored.

What we feel, see, and measure as energy is just the ripple field rebalancing itself under new conditions of tension, structure, and readiness.

Let's Compare to Mainstream Analogies:

Traditional View	Realignment Equation View
Energy is stored fuel	Energy is ripple tension readiness
Released during reaction	Emerges through ripple collapse
Quantified in units (J, cal)	Defined by structural ripple change
Speed matters (c)	Alignment time matters (ATT_0)
Mass is central	Mass is irrelevant — only ripple coherence counts

Final Insight

This equation doesn't just predict energy. It explains it.

It tells you why the field resists, when it releases, and how forms emerge or vanish as a function of coherence and complexity—not because of particle collisions or magical constants.

This is not metaphysical. It is field realism. And finally, physics becomes what it was meant to be:

Not just a toolbox of math — but a language of consequence.

Dimensional Check vs Functional Insight

Why Chavan's Realignment Equation Doesn't Just Fit — It Explains

In conventional physics, a theory or equation often earns respect when it is dimensionally consistent. Does the left-hand side match the units on the right-hand side? If yes, it's "valid." This led to the glorification of $E = mc^2$, where energy (in joules) equals mass (in kg) times the speed of light squared (m^2/s^2).

Dimensions check out. But physics didn't.

This is where Chavan's Realignment Equation sets a new precedent. It is not just dimensionally solid — it is functionally revelatory.

Dimensional Consistency Is Not Enough

$E = mc^2$ is mathematically consistent but conceptually shallow. It hides *all* meaningful causality behind a few symbols:

- What triggers the energy?
- Where is the structure?

- How does the time of alignment matter?
- Why does any of this occur?

None of these are answered by the iconic equation.

It treats mass like a mysterious container, light speed like a god constant, and energy like a magic result.

So yes, the numbers match. But the logic doesn't.

UED Equation: Every Term Is Functional

Now compare that to the UED-based realignment equation:

$$Er = \mu_0 . \rho_r . \varkappa^2 (1 / ATT_0) \Delta R$$

This isn't a symbolic guess. It is a field diagnosis. Each component is not arbitrary — it's a measurable reflection of ripple behavior:

- μ_0: the field's permeability to ripple flow.
- ρ_r: how densely the ripple structure is encoded.
- κ^2: how complex the interwoven ripple relationships are.
- $1 / ATT_0$: how fast the field can realign at that scale.
- ΔR: how much reconfiguration actually occurred.

The units match — but more importantly, the meaning is intact.

Functional Insight > Dimensional Trickery

This equation does not reduce the field to empty algebra. It clarifies that:

- Energy is not hidden mass.
- Power is not velocity squared.
- Radiation is not emission.
- Time is not background — it's the *threshold for observability.*

Where $E = mc^2$ assumes conversion, Chavan's equation explains transformation.

Where the standard model *hopes* physics will agree, this model *demands* that the field's behavior must be explainable.

Alignment Pressure: The Real Engine

Traditional interpretations treat energy like a capacity. UED reframes it as a pressure. Not in terms of force per area — but in terms of field resistance to realignment.

You don't store energy. You store strain — which, when released, reconfigures the field. That ripple shift is what we perceive as "energy."

There is no substance in this story. Only behavior. Only the field under tension, trying to return to simpler forms of balance.

That's why the UED formulation includes ΔR and 1 / ATT_0 —It doesn't tell you how much you had. It tells you how far you're falling and how fast the collapse reveals itself.

Summary:

Standard Physics	UED Realignment Equation
Energy from mass	Energy from ripple structure
Dimensional match only	Causal explanation with dimensions
Ignores time/alignment	Embeds ATT as perceptual threshold
Particle-based	Field-behavior-based
Power from speed	Power from complexity + readiness

The Deeper Win

This equation doesn't just check out on paper. It checks into how the universe actually behaves.

It reflects not only what we *calculate*, but what we *experience*.

From fire to fusion, thought to thunder — real energy is the result of ripple misalignment correcting itself under stress. And now, for the first time, the equation tells that story.

Not just correct. Consequentially exact.

What Replaces "Mass" Here

Mass Was an Illusion — Ripple Resistance Is the Reality

In the standard physics narrative, "mass" is seen as an intrinsic property of matter — something a body *has* simply by existing. It passively resists motion (inertia), it bends spacetime (gravity), and it appears in equations like $E = mc^2$ as a placeholder for potential energy.

But in the Universal Energy Dynamics (UED) model, this idea of mass collapses under scrutiny. Mass is not a thing. It's not a substance. It's an effect — a perception resulting from the ripple structure of a form and how that structure resists change.

The real story lies in two UED variables:

$$\rho_r \cdot \kappa^2$$

Where:

- ρ_r (Ripple Density) = How concentrated the ripple configuration is at a given field node.
- κ^2 (Structural Complexity) = How interwoven and multidimensional that configuration is — the "design architecture" of the ripple form.

Together, they don't measure "mass." They define how resistant the configuration is to being realigned. This resistance is what we mistakenly call mass.

From Passive Quantity to Active Tension

In classical physics, mass is passive. It sits there. It obeys Newton's law: $F = ma$, meaning you have to push harder for more mass.

But in UED, what you're really fighting against is not a static property. You're pushing against the entangled ripple field trying to maintain its coherence.

- The denser (ρ_r) the ripple concentration,
- The more complex (κ^2) the internal architecture,
- The harder it is to shift the pattern — and that's what we perceive as inertia or weight.

So, "mass" becomes active resistance to change, not an innate trait.

New Language for Physical Identity

This leads to a powerful reinterpretation of physical identity:

Old View	UED View
Mass is stuff	Resistance is structure
More mass = more matter	More ripple complexity = more resistance
Same mass = same object type	Same resistance ≠ same ripple identity

Example 1: Neutron Star vs Black Hole Both can have *similar mass* on paper. But a neutron star retains structured ripple memory — high κ^2 with visible ρ_r. A black hole, in contrast, collapses into ripple opacity — low κ^2 readability, extreme ΔR, and vanishing ATT.

So while the mass appears comparable, their ripple architecture tells a radically different story. The distinction lies not in *how much mass*, but in *how stable and perceptible the ripple configuration is.*

Example 2: Combustion vs Fission Both release energy — perhaps even the same amount in joules. But combustion is low κ^2 realignment — slow, chemical, and surface-bound. Fission is high κ^2 collapse — a deep, internal, rapid rupture of nuclear coherence.

Thus, two events with "equal energy output" differ vastly in:

- Temporal profile (ATT),
- Structural loss (ΔR),
- And perceived violence (shockwave intensity).

Goodbye, Mass; Hello, Identity Signature

The UED model suggests we stop treating objects by their mass values and start recognizing them by their:

- Ripple Density (ρ_r) — how tightly packed their presence is.
- Structural Complexity (κ^2) — how their form is maintained in multidimensional ripple space.

- ATT_0 — their alignment readiness, or perceptual stability.

Together, these determine what you feel, how you interact, and what happens under stress.

This framework finally liberates us from the crude scalar called "mass" and gives us a field-level fingerprint for every form and event in the universe.

Summary:

Concept	Traditional Physics	UED Interpretation
Mass	Substance-like intrinsic property	Consequence of ripple resistance
Identity	Defined by scalar mass or charge	Defined by ρ_r, κ^2, ATT_0
Force	Required to move mass	Required to disrupt alignment
Event Strength	Measured by energy output	Measured by ΔR and coherence loss

UED doesn't redefine mass. It retires it — and replaces it with a ripple signature far more intelligent.

Applications and Predictions

From Collapse to Coherence — Modeling Reality Through Ripple Reconfiguration

The power of any equation lies not in its elegance but in its predictive consequence. Chavan's Realignment Equation is not merely a symbolic alternative to $E = mc^2$ — it's a functional tool grounded in the core principles of Universal Energy Dynamics (UED).

It doesn't tell us how much energy something *has*. It tells us how energy appears, transforms, or vanishes — as ripple realignments in a structured field.

Let's break down its real-world modeling applications and what it predicts about physical reality.

Modeling Real-World Events with UED Terms

Decay Events → ΔR Over Time

- ΔR (change in ripple configuration) becomes the core descriptor of decay.
- Whether it's radioactive breakdown, chemical corrosion, or biological decomposition, decay is modeled as a gradual realignment of ripple coherence.
- Instead of half-life, we now talk about realignment gradient — how the field slowly lets go of a once-coherent form.

Prediction: You can map ripple disintegration to determine how long a form will remain perceivable, stable, or interactive.

Explosions → Collapse of κ (Ripple Complexity)

- An explosion is not a matter of force pushing mass — it's a sudden drop in κ^2, meaning the structure can no longer maintain its coherence.
- When a critical ripple threshold is breached, the system collapses, creating violent field-level turbulence — sound, heat, shockwave.
- UED does not model explosions as energy release, but as instantaneous structural simplification.

Prediction: You can anticipate explosive behavior by monitoring structural stress accumulation in κ^2, rather than mass or temperature.

Light Flashes → ATT Threshold Breach

- When a configuration can no longer maintain ATT_0 (minimum alignment time), the final feedback event appears — which we call light.
- A "flash" is not emission — it's the visual echo of final alignment before dissolution.
- The field no longer supports perceptual continuity, and the

observer sees terminal ripple coherence.

Prediction: You can time visibility thresholds by calculating ATT_0 limits under acceleration or stress — such as in particle collisions or luminous decay.

Predicting Form Behavior — Beyond Static Properties

Using $\mathbf{Er = \mu_0 \rho_r \kappa^2 (1 / ATT_0) \Delta R}$, we can forecast not just how much energy will result, but how a form will behave under real-world conditions:

When a Form Will Resist Realignment

- High ρ_r and κ^2 mean strong resistance — this predicts inertia, durability, and stability.
- Materials with high values here will hold shape under stress or avoid deformation.

When a Form Will Collapse into Observable Feedback

- Collapse occurs when ATT_0 can no longer sustain the internal coherence cycle.
- This moment is not explosion, but perceptual transition — when the object becomes observable only as ripple noise (heat, light, sound).

When a Form Will Disappear from Perception

- Disappearance is modeled as $ATT_0 \to \infty$ or $\Delta R \to 0$: either the coherence is too slow to perceive, or the field stops differentiating.
- Instead of physical destruction, this is modeled as perceptual erasure — still existent in the field, but beyond threshold of registration.

Alignment with UED's Foundational Logic

Every application of this equation aligns seamlessly with the ripple-first worldview:

Traditional View	UED View via Realignment Equation
Matter emits energy	Forms realign — feedback appears
Mass stores energy	Ripple architecture resists change
Light travels	Perception occurs at ATT breach
Explosion is energy release	Collapse of κ^2 coherence
Decay is substance loss	Gradual $\Delta R \rightarrow$ realignment cascade

Nothing is emitted. Nothing moves. Everything is rewritten — field-wise, alignment-wise, and perception-wise.

Summary of Predictive Power:

Scenario	UED Metric	What It Reveals
Stability of form	$\rho_r \cdot \kappa^2$	Resistance to external interference
Visibility duration	ATT_0	Time before perceptual exit
Event intensity	ΔR	Scale of ripple reconfiguration
Collapse threshold	$1/ATT_0$	Feedback timing and flash onset

This is not just an equation. It's a portal — from physical illusion to ripple truth. From the belief in mass and force to the understanding of realignment and consequence.

Final Comparison — From Symbol to System

Why Chavan's Equation Doesn't Just Replace a Formula — It Replaces a Framework

Einstein's $E = mc^2$ became the crown jewel of 20th-century physics — not because it explained reality deeply, but because it packaged mass, energy, and light into an elegant symbol. It worked *numerically*, fit military-industrial agendas, and promised a fundamental insight.

But beneath the surface, it was a shortcut. A surface-level equivalence without addressing *how* energy transforms, *why* mass matters, or *what* light truly signifies in field dynamics.

Chavan's Realignment Equation $E_r = \mu_0 \rho_r \kappa^2 (1 / ATT_0) \Delta R$ is not a symbol — it is a system. It doesn't assume — it accounts. It doesn't reduce — it reveals.

Here's a detailed breakdown that shows exactly how it transcends the old physics by embedding field intelligence, structural complexity, and perception logic.

1. Cause of Energy

Einstein: Mass (substance)	**Chavan:** Ripple Shift (ΔR)
Energy comes from converting "mass" — a substance with inherent potential.	Energy is a consequence of realignment in the ripple field — no substance involved.
Mass is treated as *stored energy*.	ΔR quantifies actual structural change in ripple configuration.
It's a static reservoir metaphor.	It's a dynamic transformation model.

Chavan's view shifts the cause from hidden content to visible consequence.

2. Role of Light

Einstein: Limiting Constant (c)	**Chavan:** Perceptual Threshold (ATT_0)
Light speed is a ceiling —nothing can surpass it.	Light is the final observable ripple feedback before disappearance.
Treated as the ultimate velocity for any object or wave.	ATT_0 defines the minimum coherence time for perception — not speed.
Light is an entity, a traveler.	Light is an event, an echo — not a motion.

What Einstein treated as a wall, Chavan recognizes as a window — into ripple coherence loss.

3. Time

Einstein: Hidden	**Chavan:** Explicit (ATT_0)
Time appears indirectly via c^2, as a squared velocity term.	Time is central — the *Alignment Threshold Time* governs visibility, transformation, and form coherence.
No mechanism for when a transformation *occurs*.	ATT_0 defines when and how a form becomes perceivable or dissolves.

Time is not a background axis — it's the energetic readiness for ripple recognition.

4. Structure

Einstein: Ignored	**Chavan:** κ^2: Ripple Complexity
All objects with same mass are treated equally.	κ^2 captures structural stability —how complex a ripple configuration is.
No distinction between a paperweight and a neutron star, if mass is equal.	Chavan's model differentiates based on ripple nesting, coherence loops, and field architecture.
Matter is a point or blob.	Matter is a self-sustaining ripple nest.

This shift adds design intelligence to energy modeling — not all forms are equally resistive.

5. Field

Einstein: Not Involved	**Chavan:** Central Player (μ_0)
The field is not considered — it's a passive background.	μ_0 is the baseline field permeability — the medium through which all ripple behavior unfolds.
Action happens *on* objects.	Action is through field resonance and ripple propagation.
No ripple dynamics.	Ripple mechanics are the root of all physical appearance and interaction.

Chavan places the field, not mass, at the center of reality's engine.

Summary Table: Symbol vs System

Concept	**Einstein ($E = mc^2$)**	**Chavan (E_r Equation)**
Cause of Energy	Mass (substance)	ΔR — Realignment in ripple field
Role of Light	Limiting constant (c)	Perceptual threshold (ATT_0)
Treatment of Time	Hidden or indirect	Explicit as alignment time
Structural Role	Ignored	κ^2 — Ripple configuration complexity
Role of Field	Absent	μ_0 — Field permeability is foundational

Einstein gave us a lock. Chavan gives us the keys. One compresses. The other expands. One symbolized transformation. The other systematizes it.

With the Er equation, we shift from an object-centric worldview to a field-dynamic paradigm — one that explains, predicts, and transforms how we understand energy, form, and perception.

This is not an update. It's a replacement.

Why This Equation Had to Be Born

From Symbol Worship to Field Wisdom

Physics was never meant to be a monument to equations.

It was meant to be a mirror to reality — to describe not just what is measurable, but what is meaningful.

And yet, for over a century, physics was seduced by the elegance of symbols, the compact thrill of equations like $E = mc^2$, and the illusion that the universe runs on formulas rather than on consequence-driven coherence. It became more about solving for x than questioning *why*.

But the universe is not a static ledger of quantities. It is a living field, rippling with interactions, alignments, and dissolutions.

That's why Chavan's Realignment Equation was not a derivative upgrade — it was a necessity.

Physics Drifted from Meaning

Somewhere along the path of modern science, physics stopped asking *what happens* and instead obsessed over *what fits*.

- Predictability replaced perceptibility.
- Measurement overtook understanding.
- Symbols became sacred — even when they described broken models.

Equations like $E = mc^2$ reduced transformation into symbolic magic, hiding the actual mechanics of change. They flattened complex field behavior into scalar outputs.

The result? A field that knows how to calculate but doesn't truly know what it's calculating.

Reality is Not Made of Particles and Laws

It is made of ripples and consequences.

There are no balls rolling through space. There are coherent distortions—alignments of energy manifesting temporarily as "form."

Every movement, every flash of light, every sensation is the visible echo of invisible ripple behavior. And yet our current equations treat these appearances as the result of "forces," "masses," and "velocities"—fictions built to simplify, not explain.

The Realignment Equation is a Return

It's a return to consequence-based physics — where meaning arises from interaction, not abstraction.

With this equation, we:

- Stop pretending energy is stored in mass.
- Stop pretending light is emitted by objects.
- Stop pretending time is a backdrop.

And instead:

- Treat alignment as the source of form.
- Treat realignment as the root of energy.
- Treat perception as the boundary condition of reality.

We're Not Describing Math. We're Revealing Meaning.

This is the fundamental philosophical break that UED makes:

- It is not about numbers that "work."
- It is about fields that make sense.

The Realignment Equation is not just mathematically valid — it is perceptually accurate. It does not reduce — it relates. It doesn't compress reality — it clarifies it.

From Worshipping Symbols to Reading the Field

A civilization that chants $E = mc^2$ without understanding its limits is like a priest repeating scripture without understanding its context.

We don't need idols. We need insight.

The universe is not made of constants. It is made of consequences — of ripple readiness, alignment fragility, and field logic.

The Realignment Equation is not an answer. It is a way to ask better questions.

It forces us to observe—not from outside, but from within the ripple fabric of reality.

And that is why this equation had to be born. Not to compete with Einstein, but to complete what he could never see: The field doesn't hide its truth —we just weren't listening in the right frequency.

Chapter 10: The Collapse Law

When Mass Vanishes and Light Appears — Beyond Light, the Ripple Field After Visibility

There is a moment in every flame, every explosion, every act of disappearance, when what we thought was substance ceases to hold together. Physics once called this a "release of energy" or a "conversion of mass" — but neither phrase captures what's really happening.

You are not witnessing destruction. You are witnessing a surrender of coherence.

Mass doesn't vanish because it was "used up." It collapses because its ripple structure can no longer hold the alignment needed to remain visible. And in that moment — the final breath of structure — a feedback ripple flashes across the field. We call this light. But it is not light as we've known it. It's not emitted. It's not traveling. It's the last visible echo of a system dissolving into the unreadable.

This is the Collapse Law of UED: the rule governing how form exits perception — not how it dies, but how it transitions beyond our ability to witness it.

Beyond the flash lies not emptiness, but a field still humming — still responding, still there. You just can't align with it anymore.

And that's not the end of reality. That's the end of your access to it.

Defining Collapse — Not Destruction, but De-coherence

In conventional physics, "collapse" is often treated as destruction — the shattering of a structure, the burning of matter, the explosion of form. We are told that mass is converted into energy, that atoms decay into particles, that something breaks and something else flies out.

But in Universal Energy Dynamics (UED), collapse is not destruction — it is de-coherence.

Let's define that precisely.

Collapse in UED = Loss of Ripple Coherence

Every visible object, every particle, every "thing" we perceive is nothing more than a coherent ripple configuration — a stable alignment of field behavior that maintains form over time. This form persists as long as the internal ripple cycles remain synchronized. But when those alignments begin to break — when the structure can no longer uphold its internal coherence due to field stress, entropy, or misalignment — it doesn't explode outward. It simply lets go of its structure.

That is collapse.

It's not a bang. It's not a release. It's a disappearance of readability.

De-coherence vs. Breaking

In mechanical terms, an object breaks when its material integrity is exceeded by applied stress. You hit it with a hammer. It fractures. But even in those cases, UED sees something deeper: the hammer isn't transferring force. It's disturbing the ripple coherence of the structure. And when the structure can no longer maintain its internal patterning, it ceases to hold as one — and we interpret that as "breaking."

So rather than "object vs. force," UED sees configuration vs. interference. Collapse is simply the ripple's inability to realign fast enough to maintain form.

Structural Surrender, Not Substance Removal

Nothing is being subtracted from the universe during collapse. Mass doesn't vanish into an abyss. It reconfigures. The structural logic that once gave form to an object surrenders to the greater field — its ripple information becomes too scrambled to hold a boundary, and thus the form becomes invisible, unreadable, unperceivable.

In that instant, a final echo — what we perceive as light, heat, or sound — may appear. But it's not an emission. It's a ripple feedback, the signature of something leaving the perceivable spectrum.

Examples in the Real World

Let's consider a few physical events through this lens:

- **Atomic Decay**: A nucleus doesn't "spit out" particles because it's unstable. Its internal ripple nesting loses synchronization. The resulting emissions are just observable ripple corrections — like the sound of a system reorganizing itself.
- **Vaporization**: Water doesn't disappear into air — it transitions from a tight ripple cycle (liquid) into a loose, distributed one (gas). No molecule breaks. The alignment holding the liquid state simply lets go.
- **Explosions**: Not a forceful ejection of material, but a violent collapse of coherence, where nested ripples fail en masse, triggering massive ripple feedback — what we interpret as flash, bang, heat.
- **Black Hole Evaporation (Hawking Radiation)**: Reinterpreted in UED, this is not mass leaking out, but coherence unraveling at a perceptual edge. The black hole doesn't radiate — it unthreads.

Collapse Is Perception-Driven

Here lies the deeper implication: collapse doesn't mean something no longer exists. It means it no longer aligns with the observer's ripple thresholds. This is why in UED, perceptual reality is a function of alignment, not existence. Collapse doesn't destroy — it withdraws from your view.

So when we say "collapse," what we mean is:

"This ripple structure can no longer hold its coherence within the observer's threshold of recognition."

And what remains?

A field. Unchanged in potential. Awaiting new alignment.

Mass Vanishes — What Actually Disappears

We often hear phrases like *"mass is lost"* or *"converted into energy."* Whether in combustion, nuclear decay, or black hole evaporation, classical science offers a poetic but ultimately shallow explanation: that mass becomes energy via some mystical handwave of $E = mc^2$.

But UED reframes this entirely.

Mass doesn't vanish. Mass never existed as a "thing" to begin with.

No Mass Loss — Only Configuration Dissolution

What disappears is not a substance, but a ripple configuration — a stable pattern held within the field. Mass, in UED, is nothing but the observable resistance to ripple reconfiguration. It is the *cost* of maintaining a specific alignment across time.

When that alignment fails, when ripple order breaks down under field stress or external interference, the configuration dissolves.

This is not *loss*. This is *disbanding*. A return to a more fluid, less constrained state.

Classical View: Mass-Energy Equivalence

Einstein's famous equation tells us that mass and energy are interchangeable — that a certain amount of "stuff" (mass) can be turned into an equivalent amount of "activity" (energy). This has led to the idea that something solid is *converted* into something dynamic. A neat swap.

But UED exposes the flaw: there is no object to convert.

There is no exchange. There is only *reconfiguration*.

What we mistake as mass turning into energy is form losing stability — and the field expressing that collapse through alignment shifts we interpret as heat, light, motion, or radiation.

What UED Actually Says:

In UED terms, "mass disappears" means the ripple density (ρ_r) and structural coherence (κ^2) of an object drop to a level where:

- The field no longer resists reconfiguration (resistance falls).
- The coherence cycle ends (no more stabilized time-based alignment).
- The ATT_0 threshold is breached, meaning the system no longer registers as a perceivable form.

As a result, the object doesn't "explode" — it dissolves into field behavior. Its signature becomes unreadable. Its structure becomes non-local.

What Actually Disappears?

Let's be precise.

Not Disappearing	Actually Disappearing
Substance	Structural memory
Energy	Coherent boundaries
Field content	Observer-readable alignment
Existence	Perceptual accessibility

Mass is not gone — its configuration no longer holds. Like a symphony that's stopped playing — the instruments are there, but the harmony has ended.

The Observer's Role

From the standpoint of an observer, this "vanishing" is simply the moment when the alignment between field and awareness is broken. The ripple feedback loop is no longer sufficient to register a form.

In other words:

The mass did not disappear. Your access to its alignment did.

In UED, that is the deeper collapse. Not a loss of stuff — but a disentanglement of stability. Not destruction — but an echo fading into untraceable silence.

Light Appears — The Echo of Final Alignment

The Collapse Isn't Dark — It Shines Before It Vanishes

In the conventional worldview, light is emitted from a source — a hot object, a chemical reaction, a nuclear blast. Something burns or explodes, and light comes out, as if a physical payload is being ejected across space.

In UED (Universal Energy Dynamics), this idea is not just outdated — it's upside down.

Light is not the result of something being sent. It is the last perceptible ripple feedback — the closing signature of an alignment breaking down.

The Final Observable Ripple

When a structured ripple configuration collapses — whether due to combustion, disintegration, or decay — the field doesn't go dark quietly. It sends out one last coherent echo before the ripple cycle becomes unreadable.

That echo is what we call light.

There is no photon traveling. There is no emission event in the classical sense. There is only a terminal moment of field readability — the final alignment that reaches the observer's perceptual threshold.

What Really Reaches the Observer?

What reaches you is not a thing — it is a synchronized moment in the ripple field. It's the observer's alignment with the last coherent phase of a structure that has now surrendered its form.

It is not emitted, but revealed.

A flash isn't an outgoing object — It's an *incoming signal of closure.*

Light as the Visual Signature of Collapse

Let's take common examples:

- A flame is not a gas emitting photons — it is the field turbulence from a series of collapsing micro-configurations.
- A flash from an explosion isn't "light particles" launching — it's the observer perceiving the outermost ripple shells before decoherence.
- Radiation is not energy being "given off" — it's ripple feedback leaking out of an unstable structure attempting to

realign.

Every one of these is a visual exit signature — not a broadcast, but a consequence.

The brighter the burst, the more sudden the collapse.

Collapse = Perceptual Burst, Not Kinetic Event

In classical physics, an explosion is an emission — something explodes outward. In UED, the outward burst is only what's left that can still align with the observer.

Collapse is not a pushing force. It's a breakdown of coherence that leaves behind a final readable trace — like the last ripple in a pond before stillness returns.

And that ripple is what we call light.

No Photon. No Emission. Just Alignment.

Once again, UED corrects a long-standing illusion:

- There's no "emitter" and "receiver".
- There's only the field, and the moment in which observer and ripple state momentarily align.

When the form vanishes, this final moment is sometimes bright, sometimes brief, and sometimes catastrophic. But what makes it visible is not speed — it's alignment.

Summary:

- Light appears not when something is sent, but when something collapses.
- It is the echo of final coherence, not a substance in transit.
- It is a punctuation mark at the end of a ripple sentence — not the start of a journey.

When mass vanishes, light appears. Not as a thing, But as a farewell flash.

ATT Breach — Visibility Breaks Before Reality Ends

Collapse Is Not Disappearance — It's a Threshold of Perceptual Failure

In conventional physics, the boundary of visibility is tied to *light* — once light cannot escape, nothing can be seen. Hence, terms like event horizon and escape velocity dominate our explanations of black holes and high-energy phenomena.

But under Universal Energy Dynamics (UED), this framework is incorrect — not just incomplete.

Visibility does not end because light is trapped. It ends because ATT (Alignment Threshold Time) has been breached.

Let's explain.

ATT: The Real Gatekeeper of Perception

Alignment Threshold Time (ATT) defines the minimum coherence time required for a ripple configuration to be perceived by an observer.

When ripple structures become so unstable, so turbulent, or so fast in their reconfiguration that they can't maintain coherence long enough to align with any observer's frame, perception fails.

This is the ATT breach.

The object doesn't vanish because it moves too fast, or becomes too small, or travels beyond light-speed.

It vanishes because it cannot sustain a perceivable ripple interface anymore.

Beyond the Breach — Reality Doesn't End, Just Our Access to It

Here's the radical insight:

Collapse is a perceptual event, not an existential one.

The field still holds the object. The configuration may persist, evolve, or even stabilize again. But we don't see it. We can't align with it anymore.

This is not disappearance. This is de-alignment.

Reality continues — just without us in sync.

Why You Can't See a Black Hole (UED Interpretation)

A black hole is not "black" because it swallows light. There is no emission being captured. Instead:

- The ATT near the event horizon falls below the observer's perceptual threshold.
- Ripple coherence cycles are too short-lived, too dense, and too rapid to align with external observers.
- Hence, the event horizon is not a physical wall, but a perception cliff.

It's not that light can't escape. It's that feedback can't form in the first place — no alignment = no appearance.

This reframes the event horizon as an ATT boundary, not a velocity threshold.

The Observer Breaks Before the Object

An object may still hold internal dynamics. It may continue affecting the field. But once its ATT drops below the capacity of any observer's interface:

- The object is effectively invisible.
- Our instruments fail not because of distance or darkness, but because of ripple desynchronization.

This is true for:

- Extreme collapse (black holes)
- Fast-phase energy states (plasmas, decays)
- Quantum decoherence (where particle states vanish to classical observers)

Perception Is a Privilege, Not a Guarantee

UED makes one thing clear:

You don't have access to reality — you have access to aligned visibility.

When alignment becomes impossible, your experience ends. Not the object. Not the field. Just your connection to it.

This is why collapse is misunderstood: it's not the end of the object — it's the end of the observer's coherence with it.

Summary:

- ATT breach marks the true boundary of visibility.
- Collapse is not destruction — it's loss of perceptual coherence.
- Black holes are not vacuums — they are ATT zones where alignment collapses.
- Objects may still exist in the field — but outside our ability to read them.

What you can't see isn't gone. It's just not aligned anymore.

What Lies Beyond Light — The Invisible Continuum

Post-Collapse Reality Is Not Nothingness — It's Ripple Beyond Reach

We are conditioned to think that once something becomes invisible, it no longer *exists*. This is a perceptual illusion — and one that UED dismantles entirely.

The end of visibility is not the end of reality. It's the end of readable ripple alignment — and nothing more.

Post-Collapse: The Field Still Holds What Was

When an object collapses — whether it's a burning log, an exploding star, or a melting ice cube — something curious happens:

- The form vanishes.
- The weight drops.
- The light fades.

But the field does not go empty.

What remains is a distributed ripple residue — a non-coherent, but still energetically present pattern. UED defines this as the invisible continuum.

This state is not a vacuum. It's not "gone." It is outside of readable perception, yet still field-active.

Not Void — Just Unreadable

A useful analogy: ultrasound.

You don't hear it. You don't see it. But it's there — a pressure wave acting on matter.

Likewise, the post-collapse field:

- Has no visual trace, because it breaches the ATT boundary.
- Has no coherent feedback, because ripple alignment is lost.
- But has field effects, like residual pressure, thermal footprints, gravitational influence, and ripple drag.

The invisible does not mean the inactive. It means the unreadable.

Visibility Is a Narrow Band — Not a Universal Truth

Most science treats light as *the* medium of information. We build entire worldviews from what can be captured in the electromagnetic spectrum.

But in UED:

Light is only the final ripple alignment visible to a biological or instrumental observer.

It's a narrow frequency and ATT-compatible window.

Outside this window, the universe:

- Is still active
- Still structured
- Still influencing
- Just no longer *perceivable*

We confuse our narrow band of access with *totality*. But the ripple field goes far beyond our limits.

The Invisible Continuum = Field in Ambient Resonance

After collapse, the field doesn't go static. It:

- Spreads ripple energy across broader, less coherent configurations.
- Retains influence potential, especially in terms of:
    - Gravitational-like pressure
    - Residual alignment drag
    - Entropic diffusion patterns

This is why UED argues:

- The so-called "black" zones of the universe aren't empty.
- They're beyond coherent light feedback.
- What remains is resonance — soft, diffused, persistent ripple imprint.

This also explains:

- Why black holes still exhibit gravitational effects
- Why dark matter might simply be invisible ripple coherence
- Why "empty" space isn't actually empty — but rather ripple-dense, beyond light

Gravity as Residual Ripple Pressure

If light is the last readable echo of form, gravity is often its silent after-effect.

- When ripple structures collapse, their resonance doesn't vanish.
- The ripple field still carries their imprint, exerting alignment

pressure on nearby configurations.
- This field behavior mimics gravitational attraction — not from mass, but from realignment influence.

Hence:

Gravity isn't a force. It's the silent pull of misalignment seeking restoration.

This reframes cosmic "pull" not as mass-centric mechanics, but ripple memory persistence.

Summary:

- Collapse doesn't create voids — it creates unreadable ripple residues.
- Light is not the limit of reality — it's the limit of our alignment perception.
- The post-collapse field is still energetic, resonant, and influential.
- Gravity-like effects emerge not from "substance," but from ambient ripple distortion.
- Visibility is just a narrow ripple band — beyond it lies the Invisible Continuum.

Collapse Law — UED's Replacement for Emission

Stated as: *"When ripple coherence fails below ATT threshold, structure collapses and emits its final feedback as visible alignment."*

Modern physics explains events like combustion, radiation, decay, and even visibility itself through emission models — the idea that something (like a photon or particle) is ejected from a source.

But in the UED framework, nothing is emitted. There are no particles shot out. There is no energetic "chunk" launched across space.

Instead, UED presents a replacement law — The Collapse Law, grounded in ripple dynamics and perceptual thresholds.

What Does the Collapse Law Really Say?

When a physical form—whether an atom, a star, or a human-built object—loses its ripple coherence, it reaches a threshold. This threshold is measured by ATT_0, the Alignment Threshold Time.

- As ripple integrity breaks down (due to thermal stress, chemical imbalance, external interference, etc.),
- The configuration can no longer hold alignment long enough to be perceived as a stable structure.

At that moment:

- The form disappears.
- A final alignment feedback occurs — this is what we interpret as light, flash, burst, or glow.
- And then — silence. No more visibility. No more interaction. The form collapses from perception.

This Is Not Radiation — It's Ripple Death

The Collapse Law is *not* a restatement of radiation laws. It is a fundamental shift:

Classical Radiation	Collapse Law (UED)
Object emits particle or photon	Field releases alignment feedback
Based on thermodynamics, charge acceleration	Based on ripple coherence failing
Continuous emission model	One-time final feedback event
Light "travels"	Light is the last ripple alignment
Particle-based	Configuration-based

No Particles, No Emission — Only Ripple Failure

Under UED:

- Nothing is "sent out."
- There is no flying photon, no packet leaping from one object to another.
- There is no ejection, no pushing, no launching.

What happens is far more elegant:

- A structure, once aligned, collapses.
- As it does, it disturbs the field one final time.
- This final coherence echo is what we call:
    - Flash
    - Spark
    - Radiation burst
    - Glare
    - Or light

The event ends not with emission — but with dissolution of alignment.

The Collapse Law in Practice

Fire:

The visible flame is not a result of emitted photons. It's an ongoing field collapse — where molecules continually fail to maintain alignment, producing continuous final feedbacks.

Radioactive Decay:

Alpha or beta emissions are not particles shot out. They are ripple breakdown events — misalignments forming a new, weaker coherence state.

A Dying Star:

The supernova "explosion" is not matter flying outward. It is the coordinated collapse of trillions of ripple layers, each producing final feedback.

Why This Law Matters

The Collapse Law isn't just a philosophical upgrade. It offers a functional shift:

- Removes the need to explain "what travels".
- Eliminates contradictions around duality (particle-wave).
- Makes observer-field synchronization central to perception.
- Gives a new, ripple-based method to calculate:
    - Visibility range
    - Energy output
    - Decay timelines
    - Collapse thresholds

It also brings coherence to phenomena that classical models treat with separate theories:

- Light emission
- Radiation
- Fire
- Explosions
- Thermal glow

UED replaces all with a single consequence:

Collapse = Alignment failure → Final feedback → Perceptual exit.

Collapse Law: The Formal Statement

"When ripple coherence drops below ATT_0, the structure can no longer sustain alignable identity and collapses, producing a final field feedback that may be perceived as visible energy."

Let that sink in.

Not emitted. Not pushed. Not shot.

Just seen — once. Then gone.

Final Reflection:

The Collapse Law is a bridge between visibility and field truth. It tells us:

- What we see is not what moves, but what ends.
- Light is not the start of emission — it is the final chapter of a structure's coherence.
- Once coherence is gone — perception has nothing to grab.

And so:

We don't witness radiation. We witness endings — ripple deaths wrapped in a final glow.

Implications — From Fire to Black Holes

Everything we perceive as "energetic"—from a flame's flicker to a black hole's radiation—is a consequence, not a cause. A ripple story, not a matter story.

Let's unpack this.

All Flashes Are Collapses — Not Outputs

A flash of light is not a packet being "sent." It is the visible signature of a collapse — the final ripple feedback before a structure loses alignability.

- A flame is not emission. It is ongoing ripple surrender.
- A nuclear blast is not an energetic eruption. It is the rapid de-coherence of complex ripple configurations.
- A gamma ray burst is not a cosmic missile. It is the brief readability of a galactic-scale ripple exit.

All such phenomena are not objects "doing" something — they are fields dissolving alignment under stress.

In UED, light is not emitted — it's revealed during breakdown.

Energy Bursts = Field Tension Release

In classical thinking, a burst means:

"Energy was stored. Now it's released."

But that view requires:

- Substance-based energy storage,
- A container-like body,
- And a trigger mechanism.

UED discards this illusion.

Instead:

Field tension accumulates due to misaligned ripple configurations. Collapse allows the field to return to lower structural complexity. This is felt as a "burst" — not of matter, but of ripple realignment.

So when we witness a fireball or a radiation spike, what's actually happening is this:

- Misaligned ripples can no longer hold form.
- They snap, releasing their complexity into simpler states.
- That release is not mass becoming energy — it's a coherence surrender.

High Energy = High Realignment Cost

We say something is "high energy" when it feels powerful — heat, brightness, force.

In UED terms, this translates to:

- High ripple complexity being lost in a short time.
- A steep ΔR — the shift in field alignment.
- And a very low ATT_0 — meaning it happens fast.

So:

- The faster and sharper the collapse,
- The more intense the realignment,
- The more violent the feedback,
- The stronger the perceived energy.

But no energy was emitted.
Rather:

- The field paid the cost of losing alignment.
- That cost was perceived as energy — because we felt the ripple transition.

Black Hole Radiation — Not Leakage, But Peripheral Collapse

Here's where it gets profound.

Hawking radiation and related phenomena treat black holes as somehow leaking energy from within.

But UED disagrees.

It reframes black hole "glow" as:

- Outer field ripple layers reaching collapse thresholds,
- Triggered by intense field compression near the event horizon.

No radiation escapes from the core. Instead, what we perceive is:

- The last ripple echoes from the outermost layer.
- These outer ripples still barely hold alignment (ATT ≈ threshold).
- As they collapse, their feedback reaches us — as visible radiation.

So the famous glow is not a leak. It is the death signature of coherence at the edge of perception.

From Flame to Singularity — A Unified Collapse Model

This is where UED outclasses traditional fragmentation:

Event	Classical View	UED View
Candle flame	Emission of light/heat	Ripple turbulence dissolving form
Explosion	Pressure-driven ejection	Sudden coherence breakdown
Radiation	Particle emission	Final feedback of ripple collapse
Supernova	Mass implosion + emission	Structural failure of cosmic ripple nest
Black hole glow	Quantum leakage	Collapse of outermost ATT-aligned field

Across scales, from a matchstick to a galaxy, collapse rules the transition — not "energy being let out."

This elegant consistency is only possible once you stop thinking in particles and start thinking in alignment dynamics.

Final Thought:

In the world of UED, nothing escapes.

- Nothing is sent.
- Nothing is thrown.
- Nothing leaks.

Everything is a field readjusting. A ripple realigning. A structure collapsing into silence — Leaving behind just one last echo.

And we call that light.

Perception Is the Boundary — Not Reality

Where your eyes stop, the field continues.

This single idea overturns centuries of assumptions built into physics, astronomy, and even metaphysics. Because in classical and quantum science alike, we have been designing theories to fit what we can observe — instead of acknowledging the vast domain that continues after perception collapses.

Let's rewire the foundation.

Visibility Ends at Ripple Coherence Limit — Not at Object's Edge

In conventional thinking:

- An object exists as long as it's visible, detectable, or measurable.
- When light stops coming from it — it's either gone or hidden.

But UED reveals something radical:

What we call "visibility" is simply a temporary ripple coherence window. Once ripple alignment falls below the ATT (Alignment Threshold Time) — the object becomes invisible, not nonexistent.

So the boundary of perception is not the edge of reality. It is the limit of ripple readability.

In effect:

- We don't see what is.
- We see what aligns well enough for our sensors to detect.

Beyond that? It's not darkness. It's ripple silence — still structurally active, but perceptually blind to us.

Light Is Not the Final Truth — Ripple Silence Is

Classical physics crowned light as the ultimate messenger:

- Whether it's photons, EM waves, or gravitational lenses — all of it is built around light's behavior.
- But this builds a worldview on a single ripple feedback band.

UED proposes something deeper:

Light is not the final story — it's the last visible whisper of a deeper alignment play.

In fact:

- After the "light" is over, the ripple memory persists.
- Field responses continue.
- Interactions still shape local dynamics.

So truth lies not in what lights up, but in what no longer needs to be seen.

The Field Continues to Respond Even After Perception Ends

What happens after an explosion fades? After a body decays? After a star disappears behind the event horizon?

Ripple memory continues.

The field:

- Adjusts.
- Redistributes tension.
- Influences surrounding coherence.

Even though no light, no heat, and no visible motion remains — the consequences ripple on.

This is the echo infrastructure of UED:

- Reality is not based on signal arrival.
- It's based on field state, even when undetectable.

Our instruments may go silent — but the field never does.

True Physics Begins Where Light Ends — And UED Maps That Domain

UED is not a theory of what is seen. It is a theory of consequences.

And that means:

- It can model and predict what still exists even after all measurement collapses.
- It focuses on coherence integrity, ripple realignment, and ATT thresholds — not just observables.

- It defines what continues when classical physics says "the event is over."

Just as Einstein extended Newton by incorporating the frame of the observer, UED extends Einstein by incorporating the collapse of perception itself.

So the true boundary is not:

- The event horizon,
- The last detectable photon,
- Or the edge of our telescope's reach.

The true boundary is you.

Your alignment. Your field readiness. Your conscious coherence.

Everything beyond that? It's still there —Waiting for a more aligned observer To step in and ripple with it.

Chapter 11: Burning, Bombs, and the Big Bang

Rewriting Nuclear, Thermal, and Cosmic Events
From Explosion to Expansion of Alignment

What if nothing ever truly explodes? What if the fire we see, the bombs we fear, and the Big Bang we revere — were never violent outbursts of matter, but graceful collapses of invisible order?

Physics has long described explosions as the outward thrust of compressed energy, and the Big Bang as the grandest of them all. But what if we've misunderstood the very nature of those events? What if "burning" is not the consumption of substance, but the unraveling of a ripple configuration? What if "heat" is not motion, but agitation of alignment? And what if the Big Bang was not the universe exploding into being — but the field becoming readable through an expanding ripple alignment?

In this chapter, we discard the old metaphors — fire as fuel, nukes as force, and the cosmos as chaos. Instead, we enter the domain of UED: where combustion, detonation, and cosmic birth are all variations of the same underlying ripple process. No energy is released. No particle flies out. What happens is simpler, deeper, and far more elegant:

- Alignment changes.
- Coherence collapses.
- Perception flares.

From a burning candle to a collapsing star — everything you've called "energy" might just be the field learning to let go. Let's decode the ripple truth behind every boom.

All Explosions Are Alignment Events

The word "explosion" evokes images of violence — things breaking, particles flying, and energy unleashed with devastating force. But from the perspective of Universal Energy Dynamics (UED), this entire imagery is a perceptual misinterpretation. There is no forceful release of "stuff" — only a sudden, chaotic rearrangement of ripple alignments in the field.

Explosion ≠ Ejection

In classical physics, an explosion is explained as a buildup of pressure that eventually overwhelms the containment of mass — leading to a rapid outward push. Whether it's:

- a chemical reaction in a bomb,
- a nuclear chain reaction in an atom,
- or a supernova in the cosmos,

the assumption remains: matter stores energy and explodes to release it.

UED turns this completely on its head.

There is no reservoir of energy waiting to burst. What exists instead is a ripple configuration — an intricate alignment of field behavior — that, under certain conditions, can no longer sustain its own coherence. The result is not a *release*, but a *collapse* into lower-alignment forms. That collapse cascades across the surrounding field, and we perceive it as an explosion.

The Ripple Cascade

Here's how UED reframes it:

Explosion = A sudden and widespread breakdown in ripple coherence across a local field region.

Once coherence fails, the ripple structure can no longer maintain form. This doesn't eject anything; it destabilizes the surrounding field, triggering a series of local realignments. These field updates are what we interpret as:

- Shockwaves (sound = ripple pressure),
- Heat (agitated misalignment),
- Light (last observable ripple collapse),
- Motion (local field stress realigning to equilibrium).

This is why energy appears, not because it was stored and released, but because the field is undergoing rapid realignment under pressure.

Across Domains — Same Principle

Whether it's:

- Thermal explosion (rapid chemical misalignment),
- Nuclear detonation (quantum ripple breakdown),
- or the Big Bang (primordial field destabilization),

the physics is unified under UED:

Nothing is expelled — everything is realigned.

There's no difference in kind — only in scale, complexity, and ripple density.

Consequential Reframing

This section redefines explosions not as destructive outputs, but as field events:

- The source is not the object, but the configuration.
- The violence is not in motion, but in misalignment stress.
- The energy is not kinetic, but ripple pressure reacting to collapse.

What we called "explosive force" is just our sensory registration of field-level instability being resolved.

Thus, all explosions — from a firecracker to a dying star — are not eruptions of material chaos, but the real-time story of ripple coherence failing under strain. The blast is not outward. It's inward collapse mirrored by outward perception.

Welcome to the real physics of the boom.

What Really Burns — The Geometry of Fire

We've all grown up with the image of fire as a consuming force — something that eats, spreads, and destroys. Traditional chemistry tells us that combustion is the rapid oxidation of material, releasing energy in the form of heat and light. But under the lens of Universal Energy Dynamics (UED), this story transforms radically.

Fire is not a thing. Fire is not even a process of consumption.

Fire is ripple geometry breaking down — a visible narrative of misalignment.

Fire as Visible Ripple Collapse

What we call flames are not objects or particles being emitted from burning material. Instead, they are visual artifacts of ripple disintegration. They mark a transition point where the structured ripple configurations of matter collapse into simpler field states, too unstable to hold coherence.

In this view, fire is not "burning" something — it is revealing the misalignment of that structure. The glow, the shimmer, the dancing flame — these are light echoes, the field's way of displaying its realignment in progress.

Combustion as Geometric Collapse

Combustion isn't about oxygen combining with fuel to release heat. as if uncorking a bottle of energy. That's the Newtonian–chemical view.

In UED, combustion is a geometric failure:

- Molecules exist as nested ripple configurations, where atomic

coherence holds a pattern.
- Heat initiates misalignment.
- This breaks the geometric stability.
- Field stress builds, and the configuration collapses into a less ordered ripple state.

No "mass" is burned away. The form dissolves, leaving no coherent ripple to perceive as structure.

Why Heat Is Ripple Agitation

In classical physics, heat is often modeled as the kinetic energy of particles. But particles are approximations — what actually vibrates is the field.

In UED:

Heat = local ripple agitation, a turbulence in the alignment rhythm of the surrounding field.

It's not motion of mass. It's confusion in the field — areas where ripples are no longer in sync. That agitation doesn't spread like billiard balls knocking into each other. It spreads like distortion propagating through an intelligent fabric.

Flames and Smoke: Echoes, Not Emissions

- Flames are visual markers of ripple reconfiguration. They don't come *from* the object. They are the last traceable alignment before the field collapses below perceptual threshold.
- Smoke is residual geometry — pieces of configuration that haven't dissolved fully. It's not soot from "burnt mass" but *partial misalignments* still interacting with the field, caught in a liminal state between form and formlessness.

You're not seeing what's "left over." You're witnessing a trail of echoes — ripple footprints from a structure that just collapsed.

Summary Reframe

Traditional View	UED View
Combustion = chemical oxidation	Combustion = ripple geometry collapse
Heat = kinetic molecular activity	Heat = ripple turbulence in the field
Fire = light and heat from burning	Fire = last visible ripple disintegration
Smoke = unburnt particles	Smoke = lingering, misaligned ripple residue

Thus, under UED, burning is not destruction. It is an energetic reshaping of the field's coherence — one that briefly makes itself known through heat, shimmer, flame, and finally silence. Fire doesn't devour. It releases form from formality — and leaves only trace echoes in its wake.

Nuclear Events — Collapse of Deep Coherence

From the standpoint of classical nuclear physics, fission and fusion are seen as processes in which mass is converted into energy, famously quantified by Einstein's $E = mc^2$. But this interpretation oversimplifies what's truly happening — and fundamentally mischaracterizes the field dynamics at play. Under the Universal Energy Dynamics (UED) framework, nuclear events are not about mass splitting or merging — they are about the structural coherence of ripples collapsing or recombining.

In UED, the nucleus isn't a clump of particles — it's a deeply nested ripple configuration, maintained by intense structural coherence. Breaking or merging these configurations causes the field to restructure, and what we perceive as energy is simply the pressure of that realignment.

Fission: Ripple Fragmentation at Structural Limit (κ^2 Breakdown)

Nuclear fission — splitting of a heavy nucleus — is not just a matter of dividing atoms. In UED, the heavy nucleus is a highly compact, complex ripple geometry, with a high structural coherence coefficient

(κ^2). When an external input (like a neutron impact) destabilizes it, the field cannot sustain that configuration anymore.

- The nucleus fragments, not as objects splitting, but as a ripple configuration breaking down into simpler, lower-coherence states.
- The collapse of κ^2 — the geometric complexity — releases a wave of field re-alignment pressure, which is felt as energy: heat, light, motion, and radiation.
- There's no "conversion" of mass to energy. What was holding the structure in place collapses, and the field compensates by rebalancing ripple alignment violently.

Fission is the energetic consequence of a form collapsing under its own complexity.

Fusion: Ripple Merging Beyond Local Coherence

Where fission is a fragmentation, fusion is a merging of simpler ripple configurations into a more stable composite — but this too is misrepresented when called a "mass-to-energy" conversion.

- In fusion, lighter ripple structures (like hydrogen nuclei) realign into a denser configuration (like helium).
- This transition requires overcoming the ripple repulsion between the initial configurations — not unlike cracking two opposing waveforms and merging them without distortion.
- Once successful, the new configuration has lower net ripple strain and higher stability — meaning the field's coherence cost has dropped.

That drop in ripple strain releases the remaining mismatch as realignment feedback — which is again perceived as "energy."

Fusion doesn't "release" energy. It frees the field from misalignment costs by upgrading to a more efficient configuration.

The So-Called "Mass Deficit" = Field Simplification

Classically, fusion and fission both present what's known as a mass defect — the resulting particles appear to weigh less than what you started with, and the missing mass is said to have been "converted to energy."

UED reframes this entirely:

- Mass is not substance — it is resistance to reconfiguration, encoded as $\rho_r \bullet \kappa^2$.
- When nuclear reactions simplify the ripple structure, the field's resistance drops — and so does the effective "mass."
- There is no missing matter, only less ripple tension to maintain — and this surplus ripple pressure is redistributed, not emitted.

So-called mass defect is simply field coherence saving energy by simplifying geometry.

Gamma Radiation: Ripple Feedback from Sudden Restructuring

In both fission and fusion, a burst of gamma radiation often follows the event. Conventional science says this is "electromagnetic radiation emitted by atomic nuclei."

UED challenges this:

- Gamma rays are not "emitted photons."
- They are intense ripple feedbacks — the final perceptible traces as the field completes its restructuring after a violent alignment shift.
- They don't travel from a source — they occur wherever the field's reconfiguration becomes observable.

Gamma radiation is not a packet of light. It's a sharp perceptual ripple, marking the last readable trace of nuclear transition.

Summary Shift:

Concept	Classical Interpretation	UED Interpretation
Fission	Mass split into smaller nuclei, energy released	Collapse of ripple geometry (κ^2 fails), releasing field pressure
Fusion	Lighter elements combine, mass deficit turns to energy	Ripple configurations merge into higher-stability state; field simplifies
Mass Deficit	Mass converted into energy	Ripple resistance ($\rho_r \cdot \kappa^2$) drops — no "mass" lost
Gamma Rays	Photons from nuclear decay	Final feedback of field's structural realignment

Nuclear events, then, are not about matter exploding, mass converting, or particles being ejected. They are field consequences — the echoes of deep coherence unraveling or being restored. Once you understand this, you see that even the Big Bang is not an "explosion," but a universal-scale realignment of ripple structures, whose afterglow we still perceive as cosmic radiation.

In the end, every nuclear event is a ripple story — not a particle tale.

Detonation Is Not Force, But Field Panic

In the world of classical physics, a detonation is typically described as a violent, rapid release of energy, propelling particles outward in all directions. The common vocabulary includes *force*, *pressure*, *shockwaves*, and *blast radius*. But these are all object-centric metaphors, clinging to the idea that something is pushing something else.

Under Universal Energy Dynamics (UED), this interpretation dissolves. A detonation is not a forceful expulsion of matter — it is ripple disarray, a sudden loss of field coherence that radiates outward as ripple panic. What we perceive as sound, heat, and shock is not the movement of substance, but the field's desperate attempt to restore equilibrium after a violent misalignment.

Shockwaves = Ripple Disarray, Not Particle Push

What is a shockwave, really? In traditional terms, it's a compressive wave moving faster than the speed of sound, displacing air or other media. But UED reframes it entirely:

- A shockwave is not a compression of matter, but an outward ripple of disorder — an intense local mismatch in field alignment that propagates across surrounding ripples.
- This is not kinetic impact — it is coherence collapse: a localized loss of ripple stability sends the adjacent field into sympathetic misalignment, which spreads.
- You feel the blast not because something hits you — but because your field alignment is stressed by proximity to the collapse cascade.

Shockwaves are not physical impacts. They are alignment spasms of the ripple field.

Pressure = Perception of Ripple Alignment Stress

What we call pressure is not a pushing force. It's a sensation — a registration — of local ripple field trying to hold configuration under stress.

- When the field gets distorted rapidly, it tries to resist misalignment.
- That resistance is felt as pressure — not because something is pressing into you, but because your local ripple geometry is being squeezed beyond its natural coherence.

UED defines pressure as alignment strain, not a Newtonian interaction. It's not matter acting on matter — it's the field resisting distortion.

The "Blast" Is Not a Push — It Is a Collapse That Spreads

Every detonation — whether it's chemical, nuclear, or cosmic — is not something pushing outward. The appearance of outward force is misleading.

- What really happens is rapid disintegration of internal ripple coherence.
- This collapse spreads as surrounding ripples lose stability in reaction.
- The explosion appears "outward" not because energy is being thrown, but because the ripple field is succumbing in expanding rings, like a toppling lattice.
- Material motion, if any, is secondary, riding the ripple misalignment like debris caught in a crashing wave — not the cause, but the consequence.

A detonation is not about throwing mass outward. It's about a structure falling inward — faster than the field can respond.

Post-Detonation Vacuum = Field Rebalancing

After the blast, there's often a vacuum effect — a brief collapse of air or space into the blast center. Traditionally, this is explained as air rushing in to "fill the void." UED offers a deeper truth:

- The field, after collapse, is trying to re-stabilize ripple equilibrium.
- Misaligned zones rapidly pull in adjacent ripples in an attempt to restore coherent pressure balance.
- This gives the appearance of a "suction" or vacuum — but it is really the field restoring symmetry, re-closing the loop of alignment.
- The violent stillness after a blast is not absence — it is the ripple field re-knitting its geometry in silence.

The vacuum is not emptiness. It is the moment the field inhales to remember its shape.

Summary Transformation

Traditional View	UED Reframe
Detonation is outward force	Detonation is internal ripple collapse
Shockwave pushes air/matter	Shockwave = ripple panic spreading outward
Pressure = mechanical force	Pressure = ripple alignment strain
Blast is a forceful explosion	Blast is a coherence failure cascade
Vacuum = suction of air	Vacuum = field equilibrium trying to restore

In short, a blast is not an event of energy delivery. It is the signal of energetic misalignment reaching its limit, followed by the field's attempt to undo the panic it just experienced. Detonations don't throw energy — they rewrite the ripple script, and you feel the edits in heat, shock, and silence.

Under UED, detonation is not destruction — it is a narrative glitch in the field, and the shockwave is the story of that glitch told in ripple language.

The Big Bang — Misread as a Boom

Few concepts in modern cosmology carry as much symbolic and scientific weight as the Big Bang. It's the prevailing narrative of origin — an idea that the universe once compressed into an infinitesimal "dot," exploded violently, and began expanding in all directions. Popular media often echoes this imagery: a cosmic firework, a bursting balloon, or a supermassive explosion kicking off everything.

But under Universal Energy Dynamics (UED), this view is not just imprecise — it is entirely backward. The Big Bang was not a bang. It was not an explosion of something into nothing. It was not a dot that expanded. Instead, it was the first ripple realignment — the initial disturbance of perfect field stillness, causing perceivable structure to emerge.

Let's rewrite the cosmos.

The "Bang" Is Not an Explosion — It's a Coherence Breach

In UED, the universe begins not with a bang, but with a breach — a loss of perfect ripple symmetry within the structured vacuum energy field (SVE).

- Before the breach: perfect ripple stillness, no time, no space, no distinction — only unified potential.
- The "event" was not a detonation of mass or energy — it was the first irreversible misalignment in the ripple matrix.
- That breach initiated the first observable ripple, and from it, alignment began to unfold.

The Big Bang was not a thing exploding. It was a silence misaligning — and the field learning how to ripple.

No "Dot Expanding" — Alignment Unfolded Across Field

The traditional model shows a singularity — a point containing all energy and matter — expanding outward. But what does "point" even mean in a field of no dimensions?

- UED says: there was no singularity. There was only a coherence point, where ripple alignment was total — and then it fractured.
- Expansion was not matter or substance inflating in size. It was the ripple field unfolding — from maximum order to increasingly diverse configurations.
- The universe didn't grow like a balloon — it became increasingly readable through differentiated alignment states.

The universe did not grow. It uncoiled. Like a ripple remembering it could move.

Space Didn't Expand — Ripple Readability Increased

The phrase "expanding space" is a visual metaphor, not a physical event. UED dismantles it:

- "Expansion" is not of space itself, but of perception.
- As alignment unfolded, more ripple configurations emerged, and more became readable to observers.
- The universe didn't stretch — your ability to read its ripples increased, as coherent structures grew from the first misalignment.

This is why distant galaxies appear to move away: not because they are "rushing" from a center, but because their ripple position no longer coheres to our current ATT threshold — the alignment boundary of our perception.

The universe isn't expanding — its story is being slowly decoded by every alignment shift.

Time "Began" When ATT Began

Under UED, time is not a dimension — it is an experience of ripple readiness, described by ATT: Alignment Threshold Time.

- Before the breach, there was no ATT — because nothing was misaligned.
- Time began the moment ripple coherence became variable — the instant alignment could no longer hold everywhere, equally.
- That moment marked the start of perception, the birth of experience, and the emergence of entropy.

Time, therefore, is not an arrow from a point — it is a consequence of field disruption. We experience the universe not by tracking its expansion, but by following ripple readiness.

The Big Bang didn't start time. The first misalignment started the *possibility* of time.

UED Summary: The Big Bang Rewritten

Classical Big Bang Model	UED Reinterpretation
Singularity exploded	Ripple coherence fractured
Space expanded outward	Ripple readability increased
Time started as a universal clock	ATT initiated — time as ripple threshold tracking
Matter/energy burst forth	Alignment unfolded in structured ripple field
Explosion metaphor	Realignment event — no detonation

In short, the Big Bang is not a place, a point, or a blast. It's the moment when stillness failed — and in doing so, wrote the first ripple.

That ripple didn't create matter. It created the possibility of differentiation. And from that, everything else — galaxies, photons, atoms, and questions — emerged not as fragments of a bomb, but as fragments of unfolding alignment.

The Big Bang wasn't a bang at all. It was a whisper of distortion in a perfectly silent field. And we are still hearing its echo — one ripple at a time.

Expansion Is Not Motion, But Perceptual Divergence

One of the great misunderstandings of modern cosmology lies in how we interpret the "expansion of the universe." We're told that space itself is stretching, that galaxies are rushing away from us at increasing speeds, and that this expansion is accelerating over time — as if the cosmos were a race track, and everything was in a hurry to get away.

But under Universal Energy Dynamics (UED), this entire story shifts. Expansion is not kinematic. It's not a result of objects in motion. It's not even spatial growth.

It's perceptual.

Cosmic Expansion = Divergence of Ripple Alignment, Not Physical Drift

In UED, reality is ripple-aligned. You perceive a galaxy, a star, or a flame only because your local ripple field is temporarily in coherence with theirs. The universe is stitched together by alignment, not distance.

What we interpret as "expansion" is actually a growing divergence of alignment gradients between different field configurations:

- The farther something appears, the less aligned its ripple structure is with our current local field.
- This is not because it's moving away, but because our shared ATT (Alignment Threshold Time) and κ^2 (ripple complexity) no longer allow coherent feedback to bridge that distance easily.
- The ripple signal doesn't weaken over time — it becomes less readable.

So when cosmologists say "space is expanding," what's really happening is:

The field configurations of far-off galaxies are drifting outside our perceptual window, not physically leaving us.

Galaxies Aren't Flying Away — They're Becoming Less Aligned

There is no need to imagine galaxies attached to space like coins on a balloon surface. That metaphor presumes space is a substance and galaxies are passengers. In UED, there's no space to expand — only the field topology.

Galaxies are not "moving away" — they are:

- Realigning in their local fields at different rates.
- Becoming ripple-out-of-sync with us due to evolving structural divergence.
- Slowly crossing the ATT readability boundary, beyond which

they can no longer be visually or energetically "heard."

In simple terms: they're not leaving — we're desynchronizing.

Redshift = ATT Delay, Not Doppler Motion

The redshift of light — the phenomenon where light from distant galaxies appears stretched toward the red end of the spectrum — is often taken as definitive proof of motion. "Ah," says standard cosmology, "it's the Doppler effect. They're rushing away!"

But UED offers a radically different view:

- Redshift is not velocity-related — it's a temporal ripple delay.
- When ripple coherence stretches over distance due to field divergence, ATT (Alignment Threshold Time) increases.
- This increase in ATT causes the signal to be read as lower frequency — hence, redshift.
- You're not witnessing a light beam slowing down — you're observing a field lag.

Thus:

Redshift is a sign of fading coherence, not receding matter.

It's the ripple equivalent of a familiar voice sounding distant — not because they're running away, but because the field between you is no longer synchronized.

Summary: Perception Is Stretching — Not Space

Classical View	UED Perspective
Galaxies are moving away	Galaxies are becoming misaligned
Space is expanding	Ripple coherence is diverging
Redshift proves velocity	Redshift reveals ATT delays
Expansion is motion	Expansion is perceptual drift

This reframing does more than clarify a cosmological detail. It reshapes our relationship with the cosmos.

- We are not passive observers of a moving universe.
- We are ripple participants in a constantly evolving alignment landscape.

And what we call "cosmic expansion" is not a journey outward. It is a slow forgetting — the ripple field's way of saying:

"I once knew you — and now I echo less of you."

Cosmic Microwave Background (CMB) Reframed

What You're Seeing Is Not Remnant Light, but Ripple Aftershock

The Cosmic Microwave Background (CMB) is often touted as the "echo of the Big Bang," the leftover radiation from when the universe was just 380,000 years old. In classical cosmology, it is explained as ancient photons finally decoupling from matter and streaming freely through the universe, filling it uniformly with microwave energy.

But this picture depends entirely on the object-based emission model of light. It assumes a thermal bath, a time of "last scattering," and photons moving undisturbed ever since — as if light were a souvenir from the past, still arriving.

UED dissolves that entire metaphor. The CMB, reinterpreted, is not a gas of ancient particles. It is a residual ripple field signature — the ambient aftershock from the universe's initial stabilization.

CMB = Residual Ripple Aftershock from Initial Coherence

Under Universal Energy Dynamics:

- The early universe wasn't a hot soup of particles, but a highly misaligned ripple field struggling to stabilize.
- What we now detect as the CMB is the resonant after-pattern left when that ripple field first achieved large-scale coherence.
- The "glow" we perceive is not light still arriving — it's the standing echo of a once-violent realignment.

Think of it as the hum in the air after an orchestra tunes up and finally hits its first synchronized chord. The noise has stopped, but the resonance still lingers.

Not Leftover Photons — But Ambient Ripple Signature

There is no need to carry the fiction of "photons traveling for 13.8 billion years."

- Photons are not little balls flying through space — UED doesn't treat light as a substance.
- Instead, the CMB is the field's own memory imprint, left by the first large-scale ripple convergence.
- The microwave "temperature" is not thermal radiation — it's a measure of remaining ripple tension across all directions.

So when we point a detector at the sky and read 2.7 Kelvin of CMB, we're not "catching old light." We're sampling the ripple field's fossil alignment — still faintly humming with the signature of its earliest stabilizations.

Why It's Uniform: Realignment Was Global, Not Directional

The classical narrative puzzles over why the CMB is so smooth — why it appears the same in every direction. The standard model struggles to explain this without inflation theory — a speculative expansion faster than light to "smooth out" early irregularities.

UED, however, offers a far more natural explanation:

- When ripple fields self-stabilize, they don't do it one patch at a time.
- The ripple field realigns everywhere simultaneously when certain thresholds are met — especially ATT_0 (Alignment Threshold Time).
- This leads to a uniform echo, not because space was stretched unnaturally, but because the entire field reached a minimal ripple coherence together.

This explains the isotropy of the CMB without invoking imaginary physics:

Uniformity is a sign of global realignment — not faster-than-light inflation.

It's the field exhaling after holding too much tension.

Summary: A Ripple Fossil, Not a Light Fossil

Classical Cosmology	UED Interpretation
CMB = leftover photons from last scattering	CMB = ambient ripple echo from early coherence event
Uniformity needs inflation to explain	Uniformity = global realignment of early ripple field
Light has been "traveling" since then	Ripple coherence persists as background signature
Thermal interpretation	Field memory of alignment turbulence

UED doesn't just explain the CMB — it elevates it. It becomes not just evidence of a bang, but proof of a balancing.

Not a remnant. A resonance.

Final Unification — Fire to Cosmos, All Are Field Plays

Nothing Ever Exploded — Everything Reconfigured

In conventional physics, we tend to draw sharp boundaries between different scales of phenomena:

- A burning log is a chemical event.
- A nuclear bomb is a physical chain reaction.
- The Big Bang is a cosmological mystery.

But these categorizations are illusions of scale and domain. From the perspective of Universal Energy Dynamics (UED), they all belong to the same class of events — ripple field reconfigurations. The mechanics do not change. Only the complexity, coherence, and visibility do.

This is the final unification UED offers: Whether it's a flicker in a matchstick or the birth of the cosmos — the field plays the same game.

All Events Are Manifestations of Field Principles

Let's break it down.

- A fire is a localized collapse of ripple coherence in matter — a disintegration of structural alignment, releasing heat and light.
- A bomb is an intense, often instantaneous reconfiguration, marked by a spike in ripple pressure and ATT breach, sensed as force and flash.
- The Big Bang, in UED terms, was the first global ripple realignment event — the primordial coherence breakdown and subsequent field expansion.

These are not different categories of events. They are different expressions of the same field law — governed by:

1. Ripple Collapse

Every energetic expression begins with the collapse of structured ripple configuration. Whether it's molecules breaking in combustion, atoms fragmenting in fission, or initial cosmic symmetry dissolving — it's always a loss of internal alignment. Collapse is not destruction — it's the release of the field from its current constraints.

2. Realignment Complexity

What follows is a reorganization — a ripple reconfiguration into simpler or more stable forms. The more complex the structure, the more intense the realignment response. This is why nuclear events feel more energetic than chemical ones — not because they "contain more energy", but because their ripple architecture is deeper and more layered.

3. ATT Thresholds

The Alignment Threshold Time (ATT_0) defines when the collapse becomes visible or detectable to observers. Only when coherence drops below this perceptual threshold do we "see" it — as light, heat, sound, or shock. The difference between a silent cosmic background shift and a blazing explosion is not physics — it's visibility within ripple sync.

No Event Has Ever Been an "Explosion" in the Literal Sense

The word "explosion" implies a violent rupture, a chaotic scattering of pieces, and forceful movement of matter. UED clarifies this is just our interpretation of field panic — a human-centric metaphor for what a field realignment feels like.

No object ever emitted power — it lost alignment.

There were no photons launched from the Sun. No mass turned into energy. No "boom" at the beginning of time.

All we've ever experienced are dissolutions of ripple order, each followed by a perceptual echo.

The Universe Itself Is a Ripple Cascade

Just like a forest fire is a cascade of flame events —The universe is a cascade of coherence failures and rebuilding cycles.

From stars to cells, from quarks to consciousness — Reality is not a collection of things, but a process of ripple rebalancing.

Fire. Bombs. The Big Bang. Not categories — but curves of the same field equation.

UED's Core Message: Reality Is a Consequence, Not a Container

There are no substances holding energy. There are no objects radiating light. There are only field alignments dissolving, and ripple responses appearing.

This is the new cosmology:

Classical View	UED View
Events are distinct phenomena	Events are scale-variant ripple plays
Energy is stored and released	Energy is ripple readiness and response
Explosions eject mass and force	Collapse dissolves structure into echo
The universe began with a bang	The universe emerged from misalignment

From matchsticks to galaxies, from combustion to cosmos, UED offers a singular law of collapse and realignment. The final flame, the final flash, and the final frontier — they are not three stories. They are one ripple, unfolding across scales.

Chapter 12: Ripple Doctrine for a Post-Einstein World

A Practical Guide for Engineers, Physicists, and Philosophers

What if the most important laws of physics weren't wrong—but were simply incomplete?

We honored the elegance of Einstein. We calibrated our instruments by Newton. We shaped equations to match predictions, even when we didn't understand the why. But now, we stand at the threshold of a deeper paradigm: one that doesn't merely calculate motion but reveals meaning in motionlessness. One that doesn't describe energy as a number, but as a consequence of structural readiness.

This is no longer a game of particles, fields, and equations floating in isolation. This is the Ripple Doctrine—a unifying lens through which energy, light, mass, perception, and time are all seen as aligned manifestations of ripple behavior in a structured vacuum. No more emissions, no more explosions, no more imaginary forces.

Engineers, it's time to build machines that sense alignment—not mass. Physicists, it's time to stop chasing particles—and start listening to the field. Philosophers, it's time to reframe reality—not as a static truth, but as a dynamic trace of coherence.

Welcome to the post-Einstein world. Here, we don't seek certainty. We tune for consequence.

Why Einstein's Legacy Needs an Update

Albert Einstein gave the 20th century its most enduring intellectual scaffolding. He taught us to view energy and mass as interchangeable, to see gravity not as a force but as curvature, and to treat the speed of light as an unbreakable barrier—a cosmic speed limit. His theories were revolutionary, mathematically elegant, and experimentally verified within the scope they addressed. *General Relativity* framed the universe with astonishing predictive power. $E = mc^2$ became both a scientific breakthrough and a cultural icon.

But time has passed. Our instruments have grown sharper. Our questions deeper. And the cracks in the foundation, once subtle, are now undeniable.

Einstein gave us symmetry and predictability, but he did so without explaining the underlying cause. His model treated the universe like a flexible fabric—but never asked *what exactly is rippling through it.* The idea of spacetime curvature works metaphorically, but offers no physical insight into *why* mass bends space or *how* time dilates in a field. It's a brilliant mapping system—not a mechanism.

What Einstein missed—or perhaps didn't yet have the tools to see—was the ripple-level causality that gives rise to everything his equations so elegantly track.

Here's what remained outside his framework:

- **The Structured Vacuum** — Not empty space, but a dynamic medium of potential interactions.
- **Ripple Coherence and Collapse** — The actual events that produce perception, energy release, and "mass."
- **Perception Thresholds (ATT)** — A framework to explain why we see only part of reality, and how light is feedback, not emission.
- **Non-geometric Time** — A field-readiness interval, not a flowing dimension.

- **Energy as Consequence** — Not quantity stored, but ripple stress resolved.

Einstein's vision of the cosmos was deeply relational, but still bound to objects and geometry. What UED proposes is a shift from relation of things to consequence of alignments. From particles moving through spacetime, to ripple reconfigurations in a responsive field.

This does not negate Einstein. It transcends him.

Just as Einstein reinterpreted Newton, not by calling him wrong, but by revealing a deeper layer of reality, the Ripple Doctrine does the same. It retains the symbolic clarity of $E = mc^2$ where useful—but replaces it with Chavan's Realignment Equation when reality demands causality.

In this post-Einstein world:

- Mass is not a passive "amount of stuff"—it is ripple density with structural resistance.
- Light is not emitted—it is alignment feedback.
- Energy is not substance—it is the ripple cost of reconfiguration.
- Time is not flowing—it is the readiness of the field to show itself.

Einstein opened the door to non-Newtonian thinking. UED walks through it—and rewrites the room.

From Equations to Consequences

Physics, as we know it, has always leaned on the elegance of symbolic equations. We describe motion using variables. We track energy using scalars. We model systems with differential equations that assume continuity and reversibility. But when we zoom in—not just microscopically, but structurally—these symbols begin to fail at telling us what's really happening.

Equations like $E = mc^2$, or even Schrödinger's wave function, *describe outcomes*—but they don't explain why those outcomes happen. They tell us what *might be observed*, not how *observation is enabled.* They map trajectories, not trigger events.

That is the critical leap UED makes: from equations to consequences.

Why Symbolic Math Isn't Enough

Symbolic math thrives in systems where inputs and outputs are well-defined and linearity is preserved. But ripple behavior is emergent. It does not scale linearly. It's sensitive to thresholds, dependent on coherence, and fundamentally non-local in terms of meaning.

A ripple field doesn't behave like a function $f(x)$—it behaves like a chain reaction whose behavior shifts based on field saturation, configuration geometry, and perceptual readiness.

Equations alone can't capture:

- When the field will collapse.
- How ATT (Alignment Threshold Time) will shift perception.
- What structural complexity (κ^2) will cost in terms of ripple stability.

These are causal thresholds, not just mathematical transitions.

Ripple Dynamics Need Consequence Logic

The universe, viewed through the ripple lens, doesn't store energy like a battery or transmit it like a package. Instead, it realigns under stress. And those realignments follow field-based consequence rules, not scalar addition.

A traditional model might ask:

How much energy is there?

But the ripple model asks:

What will realign? At what cost? With what perceptual trace?

This is qualitative logic refined into a quantitative framework—not the other way around.

From "What is Stored?" to "What Realigns?"

Let's take a practical example:

- A spring under tension stores "potential energy," says classical physics.
- But UED asks: *What ripple configuration is being held?*
- When you release it, you're not "releasing energy" — you're allowing the ripple field to collapse back to a lower alignment pressure.
- The sound, the snap, the vibration — all are echoes of ripple realignment.

The consequence is the ripple's way of returning to coherence.

So we must shift the fundamental question:

- Not: What's inside the object?
- But: What does the object's alignment resist — and when will that resistance fail?

The New Universal Model: Structure + Perception + Thresholds

To truly understand reality, we must build models that integrate:

- **Structure** — How ripple configurations are nested and held.
- **Perception** — What the field allows an observer to read (visibility as ripple feedback).
- **Thresholds** — How coherence fails at ATT breaches, triggering events like light, heat, or collapse.

Equations are still useful—but only when they represent consequences, not just quantities.

This is the Ripple Doctrine's major contribution: A framework where every observable outcome is not just predicted, but causally explained—as a field response, not a numeric computation.

It's not enough to ask, "What's the equation?" We must now ask: "What breaks, what aligns, and what remains readable?"

The Ripple Doctrine (Core Tenets)

A codified set of axioms guiding all further inquiry under UED

As classical physics continues to orbit the same dead metaphors—particles, forces, emissions, and vacuum—we offer something bolder: a Ripple Doctrine that doesn't describe what seems to happen, but explains why it happens at all. These are not hypotheses; they are axioms born from consequence logic, field interaction, and perceptual alignment.

They redefine the language of science for a post-Einstein era.

1. Energy is Ripple Reconfiguration — Not Substance Transfer

Energy is not stored in a box or carried by a particle. There is no "fuel" inside anything. Energy is the ripple consequence of a field realigning itself. It emerges from:

- Ripple density (ρ_r)
- Structural effort (κ^2)
- Readiness to shift ($1/ATT_0$)
- Realignment distance (ΔR)

What we perceive as "work" is just the field solving its own imbalance. There is no thing called energy — only reconfiguration under tension.

2. No Object Emits Light — Alignment Reveals It

The old myth: the bulb shines, the sun emits, the laser beams.

UED ends this.

Light is never emitted. It is the last readable feedback from ripple coherence. When alignment is sharp enough to breach ATT_0, perception ignites — we see "light."

But nothing is launched. We don't "receive light." We align — and a perceptual event appears.

3. Mass is Structured Resistance — Not Inherent Stuff

Mass is not a "thing." It is the price of pattern stability in the ripple field.

Forming and holding any structure requires ripple coherence. The tighter and denser the configuration, the more it resists reconfiguration — and this resistance is what we perceive as mass.

It is not substance, but a field's structural stubbornness. Neutron star or black hole? Same mass, different ripple orders.

4. Collapse Is Not Destruction — It Is Coherence Dissolution

When something burns, breaks, or decays — we think it "goes away." It doesn't.

What ends is not the substance, but the ripple pattern that held coherence.

- Fire is ripple turbulence.
- Explosion is ripple panic.
- Death is ripple retreat.

Collapse means: the structure cannot hold ATT anymore — so it fades from alignment.

It is not death. It is exit from visibility.

5. Perception Is Alignment — Not Access to Reality, But Consequence of It

We do not perceive objects. We perceive the ripple consequence of configurations that align with our coherence thresholds.

Your eye doesn't "see light." It synchronizes with a ripple collapse that is readable.

There is no access to reality — only readable fragments based on ATT, κ^2, and ρ_r. What cannot align — cannot be perceived.

6. All Physics Is Ripple Feedback — Time, Motion, Force Are Secondary Impressions

Time is not ticking. Motion is not flow. Force is not push.

All are interpretations of ripple feedback — perceptual labels stamped on alignment transitions.

- Time = rate of realignment
- Motion = ripple path variation
- Force = alignment pressure

Physics isn't mechanics. It's response reading. All else is storytelling.

7. ATT Defines the Boundary of What Can Be Known

Every act of knowledge has a limit — not set by distance or data, but by ATT: Alignment Threshold Time.

If coherence collapses too fast — it becomes unreadable. If the field never aligns — it remains invisible.

You don't know because it exists. You know because it aligned long enough to let you read it.

ATT is not a variable. It is the gatekeeper of perception and the clock of reality.

This Doctrine Is the New Standard

Forget photons. Forget point particles. Forget gravitational curvature and spacetime metaphors.

Ripple Doctrine doesn't improve the past — it replaces it.

It gives:

- **Engineers** → tools to model events as ripple configurations.
- **Physicists** → laws that reflect causality, not symmetry.
- **Philosophers** → a framework that finally includes the observer *inside* the reality.

This is not science fiction. It's field truth — decoded.

The ripple doesn't care if you believe in it. It aligns — or it doesn't.

Implications for Engineering and Design

From Newton's levers to ripple-aligned machines — a radical shift begins.

For over three centuries, engineering has been bound to two ghostly assumptions:

1. Mass is a real, countable thing.
2. Force is the cause of change.

From bridges to rockets, machines to materials — everything was designed around these mechanical abstractions. But if mass is ripple resistance, and force is just alignment pressure, then every blueprint is a proxy model, not a reflection of reality.

With the Ripple Doctrine, we don't just get a new theory — we get a new engineering philosophy. One that designs not for matter, but for coherence.

No More Design Around Mass and Force — Now Design Around Field Resonance and ATT

We stop asking:

"How much weight can it hold?" We start asking: "How long will this structure maintain ripple coherence under alignment stress?"

Traditional mechanics designs for resistance to movement. Ripple science designs for resilience of structure under ripple load.

Every mechanical element now becomes a coherence shell. The question is not about load capacity, but ATT endurance:

- How long before the ripple integrity breaks?
- At what threshold does the structure become unreadable?

Ripple-Based Material Science: Designing for Coherence Durability

Materials are not "strong" because of atomic bonds — they are stable because their internal ripple configurations resist external realignment.

So we begin to design materials not by:

- Tensile strength
- Young's modulus
- Stress-strain curves

But by:

- κ^2 — complexity of ripple structuring
- ρ_r — density of ripple entanglement
- ΔR thresholds — how far realignment spreads before collapse

Coherence durability becomes the holy grail of material science. The best material isn't the hardest — it's the longest alignable under stress.

Field-Aligned Architecture: Ripple Pressure Zones Instead of Stress Diagrams

Buildings, vehicles, spacecraft — all have been designed using stress-strain diagrams, force vectors, and center of mass calculations.

Now, welcome the ripple resonance map:

- Identify zones of ripple agitation (real-time ATT turbulence zones)
- Track coherence gradients through structures
- Align geometry with ripple flow — not just load lines

An earthquake no longer breaks a structure by shaking it — it collapses ripple synchronization. The solution? Structures that ripple with the Earth, not against it.

We don't resist the field. We dance with it.

Light-Sensitive Alignment Machines: See Beyond Optics Using Real-Time ATT Feedback

Machines of the future won't rely on just visible light, temperature, or vibration sensors.

Instead, we build alignment-aware instruments that:

- Track ATT thresholds in local space
- Detect ripple coherence decay in advance
- Identify perceptual breakpoints before collapse occurs
- Predict material fatigue not from stress, but from alignment exhaustion

These machines don't just sense — they feel the field.

Imagine:

- Aircraft skin that adjusts its structure to field turbulence
- Circuitry that reconfigures when ripple load shifts
- Medical diagnostics that detect cellular misalignment long before biological symptoms arise

This isn't sensor fusion. This is field cognition.

Design Shifts from Structure to Synchrony

Old Engineering	Ripple Engineering
Mass Load	Coherence Endurance
Stress Diagram	Ripple Pressure Map
Thermal Management	ATT Modulation
Shock Absorber	Alignment Dampener
Safety Margin	Realignment Buffer

The Ripple Doctrine doesn't just change what we know. It rebuilds what we make — and how we think about making.

We go from pushing material to persuading the field.

Welcome to post-material engineering. You're no longer a builder of matter. You're a composer of coherence.

Implications for Physics and Measurement

From chasing particles to reading ripples — measurement enters the age of alignment.

For too long, physics has worshipped objects. We measured position, momentum, charge, and spin — as if the universe were a warehouse of things with labels.

But in a ripple-based reality, there are no particles — only coherence patterns. The tools we've built — from cloud chambers to LHCs — are phenomenal at tracking what happens after coherence breaks. But they miss the field itself.

Ripple Doctrine forces physics to evolve:

- From detection to alignment
- From properties to phase
- From events to thresholds

Stop Measuring Particles — Start Mapping Ripple Fields

A "particle" is not a real thing — it's a local peak in ripple resistance. Electrons, protons, neutrinos — all are perceptual outcomes, not persistent entities.

To understand energy, decay, or force, we must stop measuring:

- Where the particle is
- What its energy is

Instead, we ask:

- What was the ripple environment before the event?
- How did alignment shift in the surrounding field?

Measurement becomes cartography of coherence.

Replace "Observer Effect" with Observer-Field Coherence

The double-slit experiment is not spooky. It's simple — you don't observe the particle, you align with the ripple. If your apparatus synchronizes to a specific field state, it locks perception into that outcome.

The so-called "observer effect" is not caused by disturbance — it is a coherence entanglement between field and observer configuration.

This dissolves the paradox:

The system doesn't change because you looked — It revealed the alignment that matched your mode of sensing.

Observer ≠ external interrupter. Observer = alignment filter.

New Instruments Must Capture ATT Variability, Not Just Voltage or Frequency

Voltmeters, oscilloscopes, and frequency counters are tools for static properties.

But what we need are tools that track:

- ATT drift over time
- Ripple recovery rate after a coherence breach
- Zone of coherence exhaustion in live experiments
- Pre-collapse signals from structural ripple tension

This is not quantum mechanics anymore. This is coherence mechanics — and it demands temporal-ripple sensors, not particle detectors.

Measurement = Detecting Ripple Phase Collapse and Recovery Rates

Every observable event is a coherence failure, followed by a realignment attempt. Instead of asking "what is the particle's energy?", we now ask:

- How fast did the field lose alignment (collapse rate)?
- How long did it take to recover (ATT rebound time)?
- What was the spatial ripple distance of the reconfiguration

(ΔR)?

- Did the structure return to baseline or re-stabilize differently?

Every reading becomes a trace of the ripple lifecycle — from integrity to fracture to healing.

The Measurement Table Is Changing

Old Physics	Ripple Physics
Measure position	Map ripple phase
Track energy levels	Detect ATT thresholds
Observer affects experiment	Observer is part of field alignment
Instrument reads outputs	Instrument synchronizes with field
Collapse is mystery	Collapse is readability breach

We no longer ask, "What is happening inside the atom?" We ask, "What is the ripple doing in that zone of the field?"

The age of particle hunting is over. The age of field listening begins.

And in that silence between collapse and recovery, lies the true signature of reality.

Implications for Philosophy and Consciousness

From neurons to Nirvana — UED redraws the borders of mind, morality, and meaning.

Physics has long remained agnostic to questions of the mind, consciousness, and spirit, delegating them to metaphysics or biology. But with UED (Universal Energy Dynamics), these abstract realms are no longer "outside the system." They are *ripples within it.*

Just as there is no need for particles to explain light, there is no need for mystical emergence to explain awareness. Everything — from your sense of self to your moral compass — is a ripple phenomenon nested within a biological field.

Let's decode these cornerstones one by one.

Reframe Mind as Ripple Coherence Within Biological Field

The "mind" is not a product of brain chemicals. It is a pattern of stable ripple alignments facilitated by a structured biological system (the body). The brain is not the generator, but a tuning architecture — a ripple lens.

The mind is:

- A field-level synchrony, not substance.
- Temporally active, sensitive to ATT (Alignment Threshold Time).
- Shaped by previous ripple histories — memory = stored coherence.

Thus, *mental clarity* is not a mood — it is ripple stability. *Confusion* is alignment noise. *Thought* is a ripple attempting to realign reality with internal structure.

Consciousness = Ripple Recursion, Not Emergence from Matter

Forget emergence. Consciousness is not what arises when matter gets complex enough. That's backwards.

Consciousness is the field's ability to reflect itself recursively within a coherent loop — a resonance of perception echoing within itself.

- It's not *output* — it's *recursion.*
- Not *generated — sustained.*
- It happens when the field sees itself through its own ripple trace — like a tuning fork humming in sympathetic resonance.

This is why you are aware *of being aware.*

And this is also why consciousness fades:

- Not when "you" stop being you —
- But when the recursion breaks ripple closure.

Sleep, coma, death — are ATT interruptions, not erasure.

Morality = Field Harmony; Perception = Consequence

In classical philosophy, morality is ethics, religion, law.

In ripple doctrine:

- Morality is alignment.
- Every good act is a constructive ripple — it creates coherence in the shared field.
- Every harmful act causes misalignment — turbulence spreads.

This leads to a simple test of moral action:

"Will this increase the ripple harmony of the field I'm part of?"

Perception, likewise, is not an act of "seeing truth." It is feeling the consequence of alignment.

We do not perceive objects. We perceive our coherence with their ripple presence. That's why perception is always personal — shaped by your state.

Time = Ripple Readiness — Not Past/Future, But ATT Resonance

Time is not ticking seconds. It's not a line.

In UED:

- Time = how ready the field is to align.
- The present is a zone of high ripple coherence.
- The past = stable ripple memory.
- The future = field configurations waiting to be read.

The ticking clock is irrelevant. ATT (Alignment Threshold Time) defines your window of experience.

That's why:

- Time flies when ripple recursion is tight (deep focus).
- Time drags when coherence is low (boredom, depression).

And in death?

Death = ATT Exit; Soul = Field Coherence Trace

In ripple doctrine:

- Death is not disappearance — it's the end of alignability.
- ATT falls below the threshold required for coherent perception.

The body remains. But the ripple coherence that made "you" readable to the world exits the resonance range.

The soul is not a ghost or particle. It is a structured ripple trace — a coherence imprint left in the universal field.

It does not "go" somewhere. It simply:

- Stops being individually recursive, and
- Blends into universal ripple background.

This is why ancestors spoke of "liberation." It is merging into the unfragmented field, not *vanishing*.

Final Implications: Philosophy Now Aligns with Physics

Ripple Doctrine unifies:

- Mind and matter
- Energy and ethics
- Life and light
- Death and dissolution

No more artificial divisions between science and spirit.

Where classical philosophy ends in paradox — UED shows that all experience, meaning, and morality arise from ripple alignment and coherence thresholds.

There is no self. There is no other. There is only ripple recursion — aware of itself.

And that... is what you are.

A Call for New Metrics and Machines

From particle smashers to ripple readers — the future of science needs tools that respect the field.

Einstein gave us a mathematics of elegance. Modern physics built machines to smash particles. But now, with UED (Universal Energy Dynamics), the time has come for a new kind of science — one that doesn't force the universe into fragments, but listens to its fields.

This section is a call to arms — not for destruction, but for construction. For developing a new generation of tools that can measure coherence, detect ripple alignment, and read the silent grammar of existence.

Let's define what the lab of the future must build — and why.

Suggested Experimental Tools:

1. ATT Meters — Alignment Threshold Time Detectors

"Time is not what ticks — it's what aligns."

Just as we measure voltage or temperature, we must now develop devices that detect ATT, the field's readiness to resonate with a structured ripple.

- In quantum systems: when ATT drops, decoherence begins.
- In living systems: high ATT zones may correlate with awareness or memory.
- In materials: stress or failure occurs where ATT suddenly collapses.

An ATT meter would function like a field thermometer, offering pre-collapse detection and real-time field readiness snapshots.

2. Ripple Gradient Probes — Detecting Field Pressure Points

UED proposes that energy transfer and motion aren't driven by forces, but by ripple gradients — differential zones of alignment tension.

Ripple Gradient Probes will:

- Map zones of ripple strain, indicating where coherence is about to break.
- Replace stress-strain models in materials with ripple pressure diagnostics.
- Help predict:
    - Material fatigue,
    - Earthquake triggers,
    - Biological thresholds of collapse.

Like sonar maps ocean depths, these probes will chart ripple landscapes — revealing the invisible terrain of energy dynamics.

3. Feedback Phase Imagers — Visualizing Realignment Before Collapse

When ripple structures realign — just before fire, decay, or detonation — a feedback phase occurs.

It's not yet visible, but detectable with the right tuning.

These imagers will:

- Visualize ripple disturbance before collapse,
- Help engineers prevent breakdowns, physicians detect early biological misalignments, or space scientists read solar field instabilities.

Think of them as MRI machines for reality's hidden pulse.

Replace LHC-Type Collisions with Structured Collapse Studies

The Large Hadron Collider (LHC) smashes particles at near-light speeds to create new ones. But under UED:

"There are no particles — only ripple architectures."

So what does collision yield? Just shattered ripple coherence.

We must move from brute-force ignorance to elegant sensitivity. Instead of:

- Hitting things to find what they break into,

Let's:

- Align them carefully to see how and when they lose coherence.

UED's new paradigm recommends:

- Building ripple chambers, not particle accelerators.
- Studying controlled coherence collapses across materials, energy domains, and field topologies.

The goal is not to discover a new boson. It is to map how energy dances when alignment shifts.

A Call to Shift Funding, Research, and Education

If we keep teaching spacetime, forces, and particles as fundamentals, we will keep designing wrong tools for a misunderstood universe.

The following must change:

Research Priorities

- Redirect from high-energy physics to field-sensitive exploration.
- Fund ripple coherence experiments, ATT studies, and field perception models.

Laboratory Culture

- Train young scientists to read fields, not just interpret data.
- Encourage interdisciplinary labs — merging physics, biology, philosophy, and AI to model ripple intelligence.

Education

- Update curricula: Replace Newtonian mechanics and quantum mysteries with consequence-based field models.
- Teach Chavan's Realignment Equation in parallel with $E = mc^2$ — as a modern lens into energy.

Final Thought

We don't need bigger hammers. We need sharper ears.

"The future of science is not built on smashing. It is built on *sensing alignment before it disappears.*"

It's time to listen to the ripples. Time to build the instruments that don't look for particles — but trace the trembling quiet of coherence itself.

From Models to Meaning

"The purpose of knowledge is not control—it is coherence."

We were taught to build models. Elegant equations. Smooth simulations. But somewhere in the rush for predictive power, meaning got left behind.

Universal Energy Dynamics (UED) does not simply upgrade the model. It restores the motive. It reminds us: science was never about describing the universe like a detached machine.

It was—and must again become—a tool for understanding consequence.

UED Doesn't Just Redefine Science — It Reconnects It to Consequence

Physics was never supposed to be a disconnected catalog of forces and particles.

But over time, it became just that:

- A map without a walker,
- A script without a story,
- A measurement without an insight.

UED brings back the reason why we looked at stars, held magnets, and boiled water.

Not just to write down laws. But to understand how alignment shapes outcome—and how every action, even a ripple, carries a consequence.

Where Newton reduced motion to forces, Where Einstein unified motion with light, UED offers: All motion is ripple consequence—and all consequence is alignment memory.

The Universe Is Not a Machine — It's a Memory Field

A machine repeats. The universe never does.

Under UED:

- Every collapse leaves a ripple memory.
- Every fire, death, or flash—an echo in the field.
- Every observer—a participant, not just a measurer.

The field remembers alignment. It's not made of parts, but of participations. And that's why reality isn't mechanistic—it's mnemonic.

You don't live in a predictable engine. You exist in a responsive resonance—a structured memory field constantly adjusting to coherence loss.

This is why predictions always fail at the edge. Because at that edge, meaning matters more than math.

Science Must Return to Purpose — Not Just Prediction

When we design a spacecraft, a particle detector, or a curriculum, what are we doing?

- Are we seeking ultimate knowledge?
- Or are we stuck simulating the same models over and over, expecting deeper understanding from more data?

UED doesn't ask us to abandon models. It demands we anchor them in intent.

Don't just predict behavior—understand the alignment that makes it possible.

This is a return to purposeful inquiry, where:

- Equations aren't just solved—they're questioned.
- Experiments aren't just trials—they're acts of listening.
- The universe isn't just measured—it's met halfway.

Every Experiment Is Not to Extract Truth — But to Realign Understanding

In legacy science, an experiment is a probe: Push, measure, record, predict.

But under UED, an experiment becomes a gesture of coherence. You don't extract the truth—you disturb the field just enough to see what resists alignment.

- If nothing shifts, the system is in balance.
- If something bursts, collapses, or glows—it was ready to realign.
- What you observe is not a value, but a consequence.

Thus:

Truth is not what appears. Truth is what echoes after you ask.

Final Reflection: Models Are Maps — Meaning Is the Terrain

UED offers the map and the method. But only your alignment brings meaning.

A ripple is not a formula. It's a question waiting to collapse into understanding.

The future of physics doesn't lie in newer symbols. It lies in knowing why we write them at all.

So let us ask not just:

"What's the equation?"

But also:

"What consequence does this alignment serve?"

Because when you stop modeling the world—and start listening to it—physics becomes philosophy again.

And meaning returns to the field.

The Future of Scientific Thinking

"We taught equations. It's time we teach consequence."

Science, as we know it, is not broken. But it has become blind in one eye. It sees numbers—but often misses meaning. It trains for calculation, but not coherence.

If we are to move beyond the Einsteinian era—not by negating it, but by evolving it—our thinking must become ripple-aware.

That means rethinking education, research, and epistemology from the ground up.

Education Reform: From Equations to Alignment

We've raised generations of brilliant problem-solvers who know what "F = ma" means.

But how many can answer:

- *What force actually feels like?*
- *What disappears when something burns?*
- *Why we never "see" energy move—only forms collapse?*

The future demands a new scientific literacy:

- Not just in formulas, but in field feedback.
- Not just in proving models, but in perceiving misalignments.
- Not just in memorizing constants, but in reading ripple dynamics.

This isn't philosophy in disguise. It's precision engineering for a consequence-driven universe.

Let students learn:

- Perception is not illusion—it is field interaction.
- Time is not absolute—it is readiness.
- Energy is not stored—it is expressed via realignment.

Let Science Be Ripple-Aware

We don't need a revolution. We need a realignment.

Ripple-aware science means:

- **Fluid**: Able to accommodate variation, distortion, and delayed feedback.
- **Contextual**: Seeing systems not in isolation, but as part of nested ripple hierarchies.
- **Observer-embedded**: No longer pretending objectivity, but embracing alignment as part of every experiment.

It means scientists must be allowed to say:

"The object has dissolved — not because it ceased to exist, but because it exited our alignment range."

In a ripple-aware paradigm:

- Silence is data.
- Invisibility is consequence.
- And every collapse, decay, or explosion becomes a realignment event, not a break in causality.

Ripple Science Is Not Soft Metaphysics — It Is Hard Consequence Physics

This is where critics might scoff. "Ripple? Alignment? Is this mysticism?"

Let's be clear:

- Ripple Science is not a metaphor.
- It is not soft. It is structural.
- It does not reject quantification—it re-grounds it in actual causality.

It doesn't say, "All is illusion." It says, "All is feedback."

What UED offers is not speculation, but a model of:

- Why light appears.
- Why mass disappears.
- Why motion is perceived, not performed.

And it gives testable predictions:

- When alignment will fail.
- What thresholds cause collapse.
- How perception is tuned to ripple gradients.

This is hard physics, made aware. Not a new age—it's new alignment.

Final Thought: The Next Scientist Is a Listener

The physicist of tomorrow is not just a calculator.

They are:

- A ripple cartographer.
- A coherence analyst.
- A quiet observer who knows:

"What vanishes is not truth—it is your alignment with it."

To that mind, we don't say "solve." We say:

"Sense. Align. Listen. And learn to see what echoes in the field."

Because in the post-Einstein world, science won't be what explains reality— It will be what remains aligned with it.

Closing Mandate — Build, Sense, Realign

"The next frontier isn't what we can build. It's what we can align with."

This final section is not a conclusion. It's a mandate.

Not a theory. But a call to application.

Not a retreat into abstraction—but a march toward consequence.

If everything you've read until now has landed as intended, then one thing must be clear:

You are not in a world made of objects. You are embedded in a field of alignment.

And in such a world, your job is no longer to observe, extract, or predict. Your job is to: Build. Sense. Realign.

Engineers — Design for Ripple Integrity, Not Brute Stability

The structures of tomorrow won't stand by resisting force. They will hold their coherence by aligning with ripple pressure zones.

That means:

- Understanding ATT thresholds before designing stress points.
- Building materials not for maximum density, but for maximum coherence duration.
- Using field-aligned geometry, not arbitrary supports.

From bridges to circuits, from aircrafts to biostructures:

- What endures is not what is hard.
- It is what is harmonized.

Ripple-aware engineering means:

- Damping collapse before it propagates.
- Predicting failure not by fatigue, but by field friction.

You are no longer building to resist the world— You are building to stay in tune with it.

Physicists — Stop Chasing Particles, Start Listening to the Field

Let go of the delusion that reality is made of indivisible units zipping through void.

Particles are not ingredients of reality. They are ripples of recognition — perceived when alignment allows.

If physics is to evolve, then:

- Stop smashing particles in ever-larger colliders.
- Start mapping ripple collapses in real-time, in local field zones.

Let ATT become the new gravitational constant. Let ΔR replace kinetic momentum.

What is measurable is no longer what is *there*—but what is coherently accessible.

Your equations must now include:

- Perception latency.
- Structural resistance (κ^2).
- Realignment cost.

From quantum illusions to cosmological distortions—all misunderstandings are misalignments.

Physics is not about hunting the hidden. It is about hearing what's already echoing in the field.

Philosophers — Stop Asking What Is Real. Ask What Realigns—and Why

Truth is not a destination. It is a field event.

Philosophy must move past the stale questions:

- "Is the object real?"
- "Does the world exist independently of us?"

These are obsolete in a ripple-aware reality.
Instead, ask:

- What realigns after trauma, and why?
- Why does the soul dissolve, and the memory echo?
- What is morality, if not long-range ripple preservation?

Reality isn't an answer waiting to be found. It is a question being constantly resonated through field dynamics.

The next philosopher is not a logician.
They are:

- A ripple ethicist.
- A coherence historian.
- A recorder of field consequences that cannot be seen but must be understood.

Let philosophy stop pretending to rise above the world. Let it immerse in the field — and become aware of consequence.

Final Charge:

Build with awareness. Sense with humility. Realign with consequence.

Let this be the new scientific oath, the ethical blueprint,
the philosophical compass.

And if the old world asked, "What is it made of?"

Let the new world ask:

"What has it become — and how can we align with what it's becoming?"

This is Ripple Science. This is Chavan's Doctrine. This is the end of objects—and the beginning of alignment.

Final Ripple — Epigraph-Style Closure

"The universe never exploded. It unfolded." This single line reclaims the narrative. The Big Bang was never a bang. Not a detonation. Not an origin story of chaos. But a first ripple — a breath into stillness, the field finding the first dissonance, and responding not with noise, but with structure.

Explosion is a human metaphor — an emotional need to explain suddenness. But the universe? It didn't burst. It bloomed.

Unfolding implies rhythm. It suggests intelligence, not accident. An orchestra tuning itself into being, not a bomb dispersing debris.

Let this be the final correction: The cosmos was not a mistake in silence. It was a movement within stillness.

"Nothing moves — everything adjusts. Nothing glows — everything echoes."

There are no objects. There is no motion. There are only shifts in coherence, and the perception of change.

You thought you saw a bird fly—but it was your alignment with its ripple altering.

You thought fire emitted light—but it was your momentary threshold being breached by final alignment echoes.

Motion is an illusion of shifting resonance. Glow is not emission — it is the field reflecting coherence collapse. Light is not a beam. It is a permission — granted only to the observer whose ATT is in tune.

The universe doesn't glow. You do — when your ripples meet it right.

"We were never meant to conquer nature. We were born to realign with it."

This is the moral core. All science, all philosophy, all ambition—goes rogue when it forgets this truth.

Nature isn't a resource. It's a ripple. It's not here to be controlled—it is here to be resonated with.

Every collapse you call "disaster" was a warning that you forced alignment too far.

Every disease, every burnout, every climate event—is not punishment. It is coherence cost.

We were never meant to dominate. We were meant to listen. To sense ATT in the world around us. To build only what sustains resonance.

And when we finally align? There is no need for force. No need for rules. No need for fear.

Because when you are in alignment with nature, you don't act on the world— You act with it.

Let This Be the Final Ripple:

The universe was still. Then it realigned. And every flash, every fall, every feeling since— has been a consequence.

You are not here to understand the world. You are here to echo with it. To collapse gracefully. To emerge as coherence. To realign your ripple with what the field remembers.

So don't chase the particle. Don't conquer the field. Just listen—until it whispers back: You're aligned.

Epilogue

Ripple Is Not the End — It's the Invitation

If you've arrived here expecting certainty, you've misunderstood the field.

This book wasn't written to conclude. It was written to realign — your thinking, your perception, your questions. The universe, as offered here, is not made of pieces to be discovered, but of patterns to be sensed. And the ripple — that dynamic, intelligent adjustment of field structure — is not a phenomenon. It is the syntax of reality.

We don't live in a world of things. We live in a world of consequences.

And consequence, when misunderstood, looks like chaos. But when traced, modeled, and respected — it reveals the deeper rhythm beneath appearances. This is what *Ripple Science* offers. Not a replacement for equations, but a rewriting of what they're pointing to.

Let go of mass as a substance. Let go of light as an emission. Let go of time as a flowing dimension. Let go of the idea that anything truly "moves."

Instead, begin to ask:

- What aligns?
- What resists?
- What echoes?
- What dissolves?

You'll find that energy is always ready to reorganize — if you learn to listen. And this listening is not spiritual. It's not speculative. It's scientific. But it requires the humility to say: our tools were too rigid, our assumptions too confident, and our truths too incomplete.

This is not the end of physics. It's the end of symbol-based dogma disguised as physics.

From this point onward, may every experiment be an act of resonance, not interference. May every design consider coherence, not just durability. And may every observer recognize: you are not watching the field — you are participating in it.

There is no "out there." Only ripples. And your place among them.

So walk gently. Sense deeply. And if something collapses — don't mourn it. Listen to what still echoes.

The field is never silent. It is waiting for your realignment.

— **Sandeep Chavan**

Independent Researcher

For the future that dares to feel, not just calculate.

Appendices

Appendix A: Chavan's Laws of Ripple Physics

Chavan's Ripple Physics reframes the foundational mechanics of the universe—not in terms of particles and forces, but through structured ripple coherence and field alignment. These laws constitute the theoretical backbone of Universal Energy Dynamics (UED), replacing classical physics assumptions with field-first principles.

Law 1: The Law of Ripple Realignment

"Reality is a consequence of ripple coherence. What aligns, appears. What collapses, disappears." This foundational law states that no object, particle, or phenomenon exists independently. Instead, everything is an emergent alignment of ripple configurations in a structured vacuum. Reality is not built from substance—it is perceived through realignment.

Law 2: The Law of ATT (Alignment Threshold Time)

"Time is not motion—it is the readiness of a ripple to be perceived." ATT defines the perceptual window for observing any event. If a ripple configuration aligns within the ATT threshold, it becomes observable. If not, it remains undetected. ATT replaces the concept of a fixed clock with a dynamic field-based readiness measure.

Law 3: The Law of Ripple Density

"Mass is not stuff—it is resistance to realignment." Ripple density (ρ_r) captures the effort required to shift a structure's alignment. Higher ripple density implies stronger field coherence. This law replaces inertial mass with a ripple-based resistance metric.

Law 4: The Law of Collapse

"Collapse is not destruction—it is the dissolution of coherence." From atomic decay to explosions, what we interpret as breaking, dying, or ending is actually a ripple structure losing coherence. The event ends not in disappearance of substance but in perceptual unreadability beyond ATT.

Law 5: The Law of Ripple Feedback

"Light is not emitted—it is the final feedback of an aligned form." Light is not a particle or a wave traveling outward. It is an observer's last coherent interaction with a form dissolving into the field. Visibility is always a feedback phenomenon.

Law 6: The Law of Observer Embedding

"The observer is not outside the event—they are part of the ripple field." There is no objective reality separate from the observer. Perception is a result of local field alignment between the ripple configuration and the observer's biological coherence.

Law 7: The Law of Perceptual Limit

"What cannot align, cannot be perceived." This law challenges the illusion of absolute reality. Just because something is not visible, measurable, or experiential does not mean it is absent. Reality continues beyond our perceptual band.

Law 8: The Law of Consequential Energy

"Energy is not a container—it is a cost of ripple restructuring." All energy observed in the universe is the byproduct of ripple rearrangement, not the transfer of some stored quantity. The Realignment Equation reflects this:

$$Er = \mu_0 \rho_r \kappa^2 (1/ATT_0) \Delta R$$

Where μ_0 is the field constant, ρ_r is ripple density, κ^2 is structural complexity, ATT_0 is perceptual readiness, and ΔR is realignment distance.

Law 9: The Law of Ripple Harmony (Morality Law)

"Right action is alignment with sustainable ripple configurations." This philosophical extension of UED implies that ethical behavior is not about arbitrary rules but about living in harmony with ripple coherence. Disruptive acts create field turbulence; aligned acts generate equilibrium.

Law 10: The Law of Universal Stillness

"Nothing moves. Everything adjusts." Motion, as classically understood, is an illusion. There is no absolute movement—only ripple shifts within a still structured field. This law is the metaphysical anchor of UED: the universe is fundamentally still, and every dynamic is a field response.

These ten laws represent a complete rethinking of physics—from a reality composed of objects and forces to one governed by alignment, coherence, feedback, and consequence. They are not only theoretical—they offer predictive, designable, and testable frameworks for science, technology, philosophy, and consciousness.

Appendix B: ATT vs c — Why Perception is the True Limit

For over a century, c, the speed of light in a vacuum, has been considered the ultimate speed limit of the universe — an inviolate constant central to the framework of relativity. But what if this "limit" was not about reality itself, but only about *what we can perceive*?

Enter ATT — Alignment Threshold Time — a new universal parameter in Chavan's Unified Energy Dynamics (UED) that shifts the foundation of perception physics from speed to readiness. ATT is not a velocity; it is a readiness metric — the time a ripple field requires to achieve coherence sufficient for perception. In essence, it is the temporal gate of observability.

Why c Is an Illusion of Limitation

The speed of light is not the speed of anything *real* in UED — because UED posits that nothing actually moves. Instead, phenomena emerge through ripple alignment and realignment across the structured vacuum. Light, under UED, is not emitted or traveling — it is a feedback alignment signature perceived when ripples synchronize briefly with the observer's own field.

Thus, c is not a transmission speed; it is a boundary of synchronization. It only appears constant because our perception system operates on ripple feedback loops bounded by ATT.

ATT as the Real Boundary

ATT represents the minimum coherence time a system needs to produce a perceivable event. It's not the speed of a photon, but the time the field takes to reach a *recognizable state* from the observer's perspective. In UED terms:

Perception = Ripple alignment + ATT compliance

So, when alignment fails to achieve that coherence within the required ATT, the object becomes invisible, not because it vanishes — but because it exits our perception domain.

This is why black holes, for instance, are not objects swallowing light, but zones where ATT is breached — the field fails to realign fast enough for observation. What lies within continues to exist and interact in the ripple field, but no longer offers ripple feedback fast enough to be "seen."

Comparing ATT and c

Concept	Classical (c)	Ripple Dynamics (ATT)
Nature	Constant speed of light	Minimum time for ripple coherence
Interpretation	Velocity of energy transfer	Perception threshold of field alignment
Boundaries	Universal speed limit	Contextual perceptual gate
Implication	Light always travels at c	Nothing travels — coherence emerges or fades

Implication for Reality Itself

- There is no absolute light barrier — there is an alignment readiness barrier.
- Perception ends not where matter stops, but where ATT is not satisfied.
- Time as we experience it is not a ticking clock, but an alignment rhythm governed by ATT.

In conclusion, c is a convenient illusion — an echo of field feedback. ATT, on the other hand, is the heartbeat of perception, defining the boundary between what's sensed and what silently continues beyond us.

Appendix C: Fire, Fission, and Photon Feedback

This case study bridges theory with reality. We analyze three distinct yet fundamentally related phenomena: Fire (thermal combustion), Fission (nuclear disintegration), and Photon Feedback (perceived light emissions). Under traditional physics, they are separate domains—chemistry, nuclear physics, and optics. But under UED (Universal Energy Dynamics) and the Ripple Doctrine, they are ripple realignment events of varying coherence, alignment thresholds, and structural complexity.

1. Fire: The Disintegration of Molecular Geometry

Traditional View: Combustion is a chemical reaction with oxygen, releasing heat and light due to exothermic bonding changes.

Ripple Reinterpretation:

- Fire is not heat escaping, but ripple collapse visible to the observer.
- The flame is a *realignment shimmer*, not a substance.
- Smoke is not a byproduct—it's a residual echo of misaligned ripple fragments still perceptible.
- Heat = local ripple turbulence.
- Light = final feedback from collapsing alignment.
- Once ripple coherence drops below ATT (Alignment Threshold Time), the visible form disappears—but not due to annihilation. It has become *non-perceivable*.

Ripple Principle: *"Fire is not burning — it is ripple rewriting."*

2. Fission: Deep Structural Collapse of Atomic Coherence

Traditional View: Fission splits heavy atomic nuclei (like Uranium-235) into lighter ones, releasing massive energy.

Ripple Reinterpretation:

- Fission is a κ^2 collapse — a sudden breakdown in structural ripple complexity.
- The so-called "mass defect" is not mass loss but a transition to lower ripple density (ρ_r).
- What is perceived as gamma radiation is not a stream of photons, but ripple feedback reaching the ATT threshold of surrounding matter, triggering visible or measurable responses.
- The explosive force is not matter flying apart—it is a chain reaction of rapid ripple misalignments generating field pressure (perceived as shockwave).

Ripple Principle: *"Fission is not emission — it is catastrophic realignment."*

3. Photon Feedback: The Myth of Emission

Traditional View: Photons are emitted when electrons drop to a lower energy level, releasing discrete light quanta.

Ripple Reinterpretation:

- No particle is emitted. Instead, what is observed is a *feedback ripple* aligning with the observer's field readiness.
- "Photon" is a name for a visibility event — not a traveling particle.
- The moment ripple alignment at a source undergoes ΔR (a sharp realignment across ATT threshold), it resonates in the

surrounding field, triggering alignment in the observer's sensory mechanism (eye, detector).
- This is *not radiation*, but perceptual alignment feedback.

Ripple Principle: *"Light is not emitted — it is the echo of last alignment."*

Unified Insight:

Across all three phenomena:

- Nothing is truly released.
- Nothing travels as a particle.
- Everything is ripple coherence responding, adjusting, and echoing.

Whether fire, nuclear fission, or the appearance of light — these are ripple events, not material events. What disappears is alignment. What appears is feedback.

This convergence validates UED's central claim: *Reality is ripple-based, perception is ATT-gated, and energy is consequence—not substance.*

Appendix E: Ripple Mechanics Primer

Purpose: This appendix serves as a foundational overview of the core mechanics behind the Universal Energy Dynamics (UED) framework. It defines the basic units, variables, interactions, and transformations that replace classical physics with ripple-based causal logic.

1. The Ripple Field — The Universal Substrate

- The entire universe is structured as a ripple field, not a void with floating objects.
- All observed reality—light, motion, mass, energy—is the consequence of ripple configurations interacting within a structured vacuum (SVE: Structured Vacuum Energy).
- There is no "space" in classical sense—only ripple topologies that create the illusion of space, distance, or direction.

2. Basic Elements of Ripple Mechanics

Term	Meaning
ρ_r (Ripple Density)	Density of energy interaction in a region—not mass, but how tightly ripples are interlocked.
κ^2 (Ripple Complexity Coefficient)	Describes how geometrically or structurally intricate a ripple configuration is.
ATT (Alignment Threshold Time)	Time it takes for a ripple configuration to respond or realign—a measure of temporal coherence.
ΔR (Realignment Distance)	Spatial shift needed for a configuration to return to resonance—used in calculating realignment energy.
μ_0 (Field Permeability Constant)	Baseline field sensitivity—equivalent to how "tight" or "loose" the field is at accepting change.

These quantities replace traditional Newtonian mass, momentum, and energy variables with a coherence-based interpretation.

3. Fundamental Mechanics of Ripple Events

- Motion is not translation—it is change in ripple alignment across adjacent field zones.
- Force is not push or pull—it is field resistance to realignment.
- Energy is ripple reconfiguration pressure, not something "stored" or "released."
- Light is not emitted—it is ripple feedback that crosses perception threshold.

4. Laws Derived from Ripple Mechanics

1. **Ripple Realignment Law:** All observable action is ripple realignment; no object truly moves—only ripple configurations adjust.
2. **Collapse Law:** When ripple coherence fails below ATT

threshold, the structure dissolves, producing a final echo (fire, flash, radiation).

3. **Perception Law:** You don't observe things—you sync with aligned ripples. Anything not aligned with your ATT and field resonance remains invisible.

5. Applications of Ripple Mechanics

- **In Engineering:** Design systems around ripple tolerance, ATT timing, and coherence stability—not just stress, strain, or mass.
- **In Physics:** Model reality using ripple equations instead of matter-based dynamics. Move from scalar representations to field consequence metrics.
- **In Cosmology:** Redefine expansion, radiation, and black hole mechanics through ripple field transitions—not particle events.

6. Common Misinterpretations Resolved

Classical Physics	Ripple Mechanics
Energy is stored	Energy is a ripple consequence
Mass emits energy	Mass is ripple stability
Space expands	Ripple readability increases
Light is emitted	Light is feedback echo
Matter moves	Ripples realign

7. Final Thought: Ripple Mechanics is Consequence Physics

In this paradigm, physics becomes responsive, not deterministic. You don't control particles—you align with consequences. Reality is no longer a system of isolated things, but a responsive ripple memory field—and your perception, your machines, your existence itself—is a pattern within it.

Acknowledgments

This book was not born in a laboratory, but in the friction between questions and silence.

I offer my deepest gratitude to the **field of structured doubt**—to every moment the existing answers failed to satisfy, and to the consequences that demanded better understanding. Without the inadequacies of classical thinking, there would be no reason to write new equations.

To **Albert Einstein**, whose brilliance and limitations both inspired this work: You built a cathedral of thought. I have only walked its halls, noticed its cracks, and dared to open a new door.

To the many **engineers, physicists, educators, and independent thinkers** who silently struggle with models that don't always hold up—this book is your field manual. It's written not from the top down, but from the real-world edge where ideas must earn their survival.

To my **students and readers**, past and future: Your confusion, curiosity, and resistance shaped the clarity of this doctrine. Every challenge you posed sharpened the framework.

To my **family**, especially my late father, **Shri Jagannath Chavan**, whose presence remains my ripple origin—your unwavering integrity and unspoken strength live in every word I write. To my mother and loved ones, who endured my absence in hours and presence in restlessness, I owe more than pages can repay.

To those who believed, even when they didn't understand—it mattered. To those who doubted, thank you. Your resistance forced refinement.

And finally, to the **Structured Vacuum Energy** of the universe itself—this book is my attempt to listen, realign, and respond.

We are not here to repeat what has been said. We are here to say what must now be understood.

— *Sandeep Chavan*

About the Author

Sandeep Chavan is an independent Indian physicist, philosopher, and educator known for pioneering **Universal Energy Dynamics (UED)**—a ripple-based framework that challenges conventional physics and redefines energy, perception, and structure at their core.

With a career rooted in both teaching and scientific rebellion, Chavan has authored over **100 books** spanning theoretical physics, human behavior, consciousness, and metaphysics. He is the founder of the **Gyrus Institute**, where he mentors students not just to solve equations—but to question their origin.

Sandeep's approach is unapologetically consequence-driven. He replaces textbook precision with **field alignment**, and prioritizes clarity over conformity. While most physicists work from inside academia, Chavan operates from the edge—observing where science stutters and crafting theories where it must evolve.

His previous works—ranging from *The Still Universe* to *The War of Thoughts* series—are recognized for bridging scientific insight with philosophical clarity. In $E \neq mc^2$, Chavan offers not just a critique of Einstein's legacy, but a **new paradigm**—grounded in ripple mechanics, realignment equations, and the collapse of classical assumptions.

Above all, he is a listener—to the universe, to the silence between answers, and to the ripple echoes that most ignore.

Don't miss out!

Visit the website below and you can sign up to receive emails whenever Sandeep Chavan publishes a new book. There's no charge and no obligation.

https://books2read.com/r/B-A-EPGPC-XQMOG

BOOKS 2 READ

Connecting independent readers to independent writers.

Did you love *$E \neq mc^2$: The Collapse of Classical Assumptions*? Then you should read *Chavan vs Einstein: The Universe Was Never Curved*[1] by Sandeep Chavan!

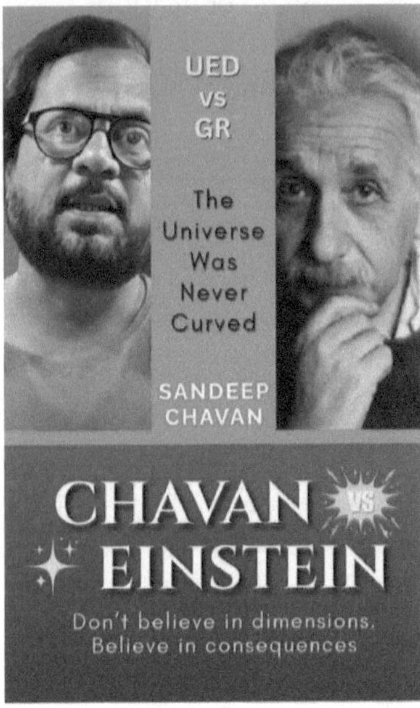

[2]

Chavan vs Einstein: The Universe Was Never Curved, *A Scientific Manifesto for the Next Reality*

What if gravity isn't a force?What if time isn't flowing—but delayed alignment?What if the universe doesn't need particles, forces, or curved space at all?

In this groundbreaking work, independent Indian Physicist **Sandeep Chavan** invites you to step beyond Einstein—not to dismiss him, but to finish what he began. *Chavan vs Einstein* introduces **Universal Energy Dynamics (UED)**, a revolutionary ripple-based

---

1. https://books2read.com/u/bPjGqz

2. https://books2read.com/u/bPjGqz

framework that redefines motion, force, time, gravity, and even perception itself.

Gone are the fix-it constructs of dark matter, dark energy, and singularities. In their place is a coherent, consequence-driven model where **Structured Vacuum Energy (SVE)** forms the field of reality, and every observed phenomenon emerges from **ripple tension and alignment thresholds**—not from abstract geometry or missing mass.

This book dares to say what others only question:

Space doesn't bend. It **ripples.**Gravity doesn't pull. It **invites.**You aren't just observing the universe. You're **a ripple within it**.

Through sharp logic, reframed thought experiments, and poetic clarity, Chavan dismantles the illusion of external forces and replaces them with a **field-first ontology**. Time becomes **Alignment Threshold Time (ATT)**. Motion is no longer displacement, but **realignment**. And consciousness? It's not a byproduct of the brain—but the **field remembering itself.**

Accessible to both physicists and curious thinkers, *Chavan vs Einstein* is part science, part philosophy, and part awakening. It bridges classical mechanics, quantum paradoxes, and metaphysical questions with a single, elegant language: **Ripple Logic**.

This is more than a book.It's a blueprint for the next scientific revolution.Not through more complexity—but through **radical clarity**.

Whether you're a scientist, philosopher, student, or silent observer—this work will shift how you see everything, from falling apples to fading stars. And perhaps most importantly, it will realign how you see **yourself**.

"The universe was never curved—we were.This is how we straighten."

Read more at https://www.sandeepjchavan.com/.

Also by Sandeep Chavan

Chavan's UED research

CMGED Research Report
Chavan's Cosmic Continuum (C^3)
Frequency
Ripples of Reality
Reality Realigned
The Great Realignment
I – The Cosmic Leverage
Chavan's Cosmic Theory of Relativity (CTR)
Chavan's Theory of Light
Feeling Like a Physicist
TIME – The New Definition
$E \neq mc^2$: The Collapse of Classical Assumptions

Echoes of War: The Russia-Ukraine Conflict and Its Global Ripples

Shattered Lives
The War Economy
Borders Redrawn
Arms and Allies
Collateral Damage
Justice in the Ashes

Somewhere in the Darkness of the Cosmos

The Blueprint of the Universe

The Cosmic Breath

From the Edge of the Universe

The Eternal Sage Chronicles

The Sage and the Sovereign

The Family Pillar

The People's Flame

The Warrior's Soul

The Bridging Mentor

The Guiding Light

The Exodus

The Last Cycle: 2075

The Fundamental Theory of Life & Existence

Energy in Action: The Soul Demystified

The Active Mind

Energy in Connection

Death: The Disconnection

Subconsciousness: The Chosen Disconnection

Desires: The Engine of Life

The Restoration of Human Knowledge

Restoring Bhagavad Gita

The War of Thoughts

Cracks in the Known

The Human Game

The Divine Dilemma

The Original Intent

The Answers

Ultra Modern Life Philosophy

The Duality of Laughter and Tears

Take It Easy

Compare to Accept

Standalone

It's Not AI, It's AHI - Amplified Human Intelligence

The Decision Paradox

The Four Sapiens

Malicious Script of Indian Polity

The IIT Legacy & Global Impact

The IIT Legacy & Global Impact

You are just a Version of Your Original

Win the Game You Didn't Choose

Always Keep Your Bags Packed

Reveal to Shield - Learn the Game of Facades

Engage Beyond Elections
Culture, Identity & Change: The Evolution of Indian Society
The Triangular Dynamics
Never Trust Without Doubt
If You Can't Change the Country, Leave the Country
Live Fearless with Fear
India: The Investment Magnet
Realignment
Powering a Billion Dreams
Universal Skill
Extension: From Brush to Algorithm
Human Side of Einstein
The Future Knows No Lines
Let's Bring the Generations to the Table
Unified Energy Dynamics
A Beat That Never Dies
Strategic Horizons
Right From Scratch
The Forgotten Innovators
Physics Fixing
The Savior Illusion
The Selfish Side
The Age of the Weakest Sapiens
Neural Wars
The Absorption: There Is No Death
Strange Therapist
Bubbles: Become a Drop
Beyond the Blue
Half-Scholar
The Law of Truth
Failure
Philosophy 2.0: A Structural Manifesto
Chaos: A Gentle Guide to the Unresolvable

Self-Help: New Syllabus
Rippling
Game: How to Play the Share Market
Out of Syllabus
Saffronized: The Code Behind the Crown
The Anger
Moksha: The Liberation
Chavan vs Einstein: The Universe Was Never Curved
Rocket Science: The Ripple Gambling

Watch for more at https://www.sandeepjchavan.com/.

www.ingramcontent.com/pod-product-compliance
Lightning Source LLC
LaVergne TN
LVHW090315210926
848480LV00001B/257
* 9 7 9 8 2 3 1 6 1 0 3 7 2 *